TABLE OF CONTENTS

Cosmogenic nuclides

Anthropogenic radionuclides

Materials sciences

Education

Publications

Talks and Posters

Seminars

Theses

Collaborations

Visitors

EDITORIAL

We hope you will enjoy reading the LIP 2016 annual report. It's our goal to present a comprehensive and easy to read snapshot of our research activities. In 2016, the last four-year funding period has ended and a new contract between our partners PSI, Empa, Eawag, and other ETH departments has been reached. This gives us great support and we will be able to continue our projects of high scientific level steam-ahead for the next four-year funding period.

It is the unique feature of our laboratory to direct not only fundamental and applied research, but also to have the technical and operational background as well as the required support to design, construct, and set into operation instrumentation at the cutting edge of present state of technology. A milestone in 2016 was certainly the start of operation of our second MICADAS system. This instrument was installed with significant financial support from the ETHZ Earth Science Department (Prof. Tim Eglinton) who is one of our main users. Now, LIP has full access to the latest generation AMS system that includes "GreenMagnets" and He gas stripping. It is remarkable that the new instrument is capable of measuring individual modern samples not only at the counting statistical limit of 1 permil, but also provides sufficient operational stability to run batches of standard reference materials with an overall reproducibility of less than 0.5 permil. All this is possible at a power consumption of less than 3 kW.

Our applied research fields cover a wide variety of exciting applications. We conduct these studies with our partners or with our external collaborators from Universities of Switzerland, Europe and overseas, from national and international research and governmental organizations as well as with commercial companies. We want to thank all of them for their confidence and support. In addition, LIP provides exceptional opportunities for Master and Ph.D. students to conduct comprehensive, self-contained research projects and to launch their academic career in a research environment interconnected between various scientific disciplines. The strong link, not only between AMS and Materials Sciences, but also between AMS and the Earth and Environmental Sciences, offers opportunities to take optimum advantage of the enhanced knowledge gained by studying the physical principles that are common to those disciplines. This has led to new discoveries, which in turn, have initiating exciting follow-up projects.

It is our mission to provide both, excellent service to our internal and external users and significant contributions to the educational program of ETH. Thanks to all LIP staff members who contributed diligently, with commitment, and with remarkable passion to our activities. Without such an excellent scientific, technical and administrative staff, the success of the laboratory would not have been possible.

Hans-Arno Synal and Marcus Christl

THE TANDEM AMS FACILITY

Operation of the 6 MV TANDEM accelerator

New DAQ for TANDEM AMS measurements

^{26}Al measurements with the gas-filled magnet

OPERATION OF THE 6 MV TANDEM ACCELERATOR

Beam time statistics

Scientific and technical staff, Laboratory of Ion Beam Physics

In 2016, the 6 MV tandem accelerator was in operation for 1145 hours, similar to previous years (Fig. 1). About 40% of the time was dedicated to AMS and 50% to Material sciences, respectively, while about 10% was used for maintenance activities and tests of the terminal voltage stability. When using the Generating Volt Meter (GVM) for stabilization we noticed instabilities of the terminal voltage up to a few permil. Thus we decided to replace the old GVM with a new module from NEC, which will be installed in January 2017.

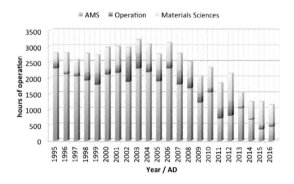

Fig. 1: *Time statistics of the TANDEM operation subdivided into AMS (blue), materials sciences and MeV-SIMS (green), and service and maintenance activities (red).*

We had two tank openings in 2016: during the first one in January we replaced the bearing blocks on the HE chain drive, replenished the carbon stripper foils (Fig. 2), cleaned and checked the column resistors. A rupture of the LE chain caused the second tank opening in June. We installed a new chain and new sheaves made out of conducting nylon. NEC has had good experience with this new material in their smaller accelerator, we are the first to test it in a 6 MV tandem accelerator. We have not reached 6 MV since the installation, but it is too early to say whether the new material is responsible for frequent sparking above 4 MV.

Due to the high voltage problems, the AMS activities were concentrated in the first half of 2016 with measurements of ^{36}Cl with 324 unknown samples. Also tests with the gas-filled magnet (GFM), which require stable operation at 6 MV, were only possible in that time. Nevertheless, first successful measurements of ^{26}Al from extraction of AlO^- with the newly built large-acceptance gas ionization chamber were done.

Fig. 2: *Used stripper foil (3 µg/cm^2). The dark shadow is from deposits of residual gas.*

The total beam time in hours used for materials science increased by about 35% compared to 2015. This is largely due to the MeV-SIMS project which used 27 days of operation mainly for reduction of the mass spectrometer background and the development of micro-imaging capabilities of organic molecules. For projects of curatorial partners and industry about 1300 samples were analyzed by the classical IBA techniques RBS, ERDA, and PIXE. This number remained approximately constant compared to the previous year.

NEW DAQ FOR TANDEM AMS MEASUREMENTS

Convenient gating and improving data analysis

C. Vockenhuber, S. Bühlmann, A.M. Müller, R. Pfenninger, H.-A. Synal

In 2015 we retired the data acquisition (DAQ) HAMSTER after almost 20 years of successful duties in the intense measurement program at the 6 MV EN Tandem. The new DAQ is based on the FASTComTec MCA3 system that is controlled by the in house SQUIRREL software, similar to our measurement setup at the TANDY. A very useful feature of the old HAMSTER was the ability for event storage, which means all the signals are recorded event by event. This allows a complete reanalysis of the dataset with new (improved) gating and regions of interest. However, due to hardware limitations the HAMSTER could record only up to 30000 events per cycle (5 to 30 s). Thus event storage could not be used in measurements with intense isobaric interference (e.g. ^{36}Cl).

AMS measurements at the Tandem require to record up to 8 signals. The FASTComTec MCA3 system is also capable of event storage and due to modern hardware event rates of several million per seconds are possible. However, the gating procedure in the MPA-NT software is cumbersome in particular when individual signals are combined (e.g. total energy as the sum of several ΔE signals or the position calculated from a left and a right anode signal). Furthermore, in a typical AMS measurement hundreds of individual runs, each consisting of 10 to 30 cycles, have to be reanalyzed. This can not be easily implemented with the MPA-NT software.

Thus we wrote a program for convenient gating and data reduction in MATHEMATICA. The program consists of three steps: First, for each measurement (magazine) the original stored data in binary format are converted into an easy readable ASCII format. Second, pre-gating is based on the individual and all combined spectra of a single run (typically of a standard, Fig. 2). Gates can be adjusted in an iterative

way, allowing to find the optimal balance between high suppression of the isobar (e.g. ^{36}S) and minimal losses of the radionuclide (e.g. ^{36}Cl). Finally the number of counts per cycle are taken from a ROI in a 2D-spectrum. In the third step the new gates are applied to all runs of the magazine and the corresponding values for the radionuclide are updated in the database.

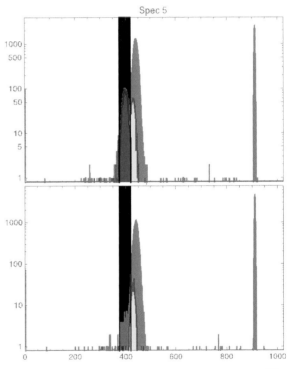

Fig. 1: *1D spectra of signal ΔE5 of a ^{36}Cl standard (top) and a blank (bottom). The grey spectra shows all counts, in red all events that passed the previous gates and in blue the events that passed gating on all signals.*

MATHEMATICA's implemented parallel computing environment allows to take full advantage of modern multi-core processors and allows to reanalyze a measurement within a few minutes.

^{26}AL MEASUREMENTS WITH THE GAS-FILLED MAGNET

Exploring high intensity isobar separation of ^{26}Mg at moderate energies

K.-U. Miltenberger, A.M. Müller, M. Suter, C. Vockenhuber, H.-A. Synal

During the last years the gas-filled magnet (GFM) installed at the end of the 6 MV EN Tandem AMS beamline could be revived due to the advent of robust silicon nitride (SiN) foils [1]. In case of ^{26}Al measurements with AlO$^-$ ions, the GFM enables suppression of the high intensity isobar ^{26}Mg and thus promises an improved overall measurement efficiency.

The GFM separates isobars based on their different mean charge state in the gas. Based on trajectory simulations of ^{26}Al and ^{26}Mg inside the GFM a new gas ionization detector (GID) with a large (30x40 mm^2) SiN entrance window and angled flange was designed and built. The five-anode configuration – including a diagonally split second anode providing a position signal – was optimized for separation of ^{26}Al and ^{26}Mg ions based on their energy loss in the detector gas [2]. To improve the electric field distribution the detector stack was positioned asymmetrically with respect to the entrance window and the cathode was shortened (Fig. 1).

With the applied gating optimized for high ^{26}Mg suppression (Fig. 2) – and thus lowering the acceptance of ^{26}Al – the ZAL94N standard was measured to ≈40% of its nominal ratio of ^{26}Al/^{27}Al = 480×10^{-12} and the AlBl blank sample was measured at a ^{26}Al/^{27}Al ratio of approximately 1.5×10^{-14}.

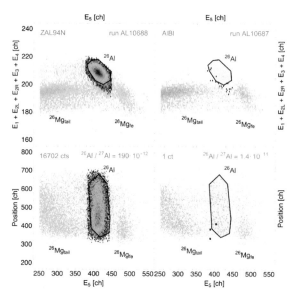

Fig. 2: *Gated 2D energy-loss spectra of a ZAL94N standard (AL10688, left) resp. an AlBl blank sample (AL10687, right).*

Despite the lower ^{26}Al acceptance, the ^{26}Mg suppression provided by combination of GFM (10^3) and GID (10^4) enables ^{26}Al measurements utilizing the significantly higher currents provided by AlO$^-$ ions. Therefore, the overall measurement efficiency is about five times improved with respect to the previous setup at the 6 MV EN Tandem accelerator.

[1] C. Vockenhuber et al., LIP Annual Report (2015) 10

[2] K.-U. Miltenberger et al., Nucl. Instr. & Meth. B, submitted

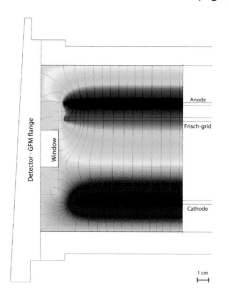

Fig. 1: *Simulated electric field distribution near the detector entrance window.*

THE TANDY AMS FACILITY

Activities on the 0.5 MV TANDY in 2016

New small Faraday Cups for the TANDY

A 300 KV multi- isotope AMS facility

The ^{41}K interference in ^{41}Ca measurements

Calcium and strontium metabolism compared

ACTIVITIES ON THE 0.5 MV TANDY SYSTEM IN 2016

Beam time and sample statistics

Scientific and technical staff, Laboratory of Ion Beam Physics

In 2016, the multi-isotope facility TANDY (Fig. 1) accumulated more than 2900 operation hours.

Fig. 1: *The inner parts of the Tandy accelerator*

The year successfully started with a second proof of concept experiment for the development of the new 300 kV multi-isotope system. For this experiment routine AMS operation of the Tandy was suspended for about eight weeks. Nevertheless, about 2400 AMS samples (here always referring to unknown samples not including standards and blanks) were analyzed over the course of 2016.

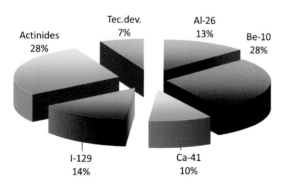

Fig. 2: *Relative distribution of the TANDY operation time for the different radionuclides and activities in 2016.*

In total, more than 90% of the beam time was dedicated to routine AMS analyses of user samples while about 10% of the beam time was

reserved for testing and developing new instrumentation (Fig. 2).

More than 1200 samples were analyzed for ^{10}Be, most of them for in-situ dating applications and ice core studies (Fig. 3). More than 500 samples were actinide (U, Pu, Am, Cm, Bk, Cf) analyses for users working in the field of oceanography and environmental monitoring, or for human bioassay studies. In this context, the 1000th ^{236}U sample was run in June 2016. Almost 400 ^{129}I samples were analyzed mainly for oceanographic and environmental monitoring applications. The numbers of ^{26}Al samples increased because routine operation was completely transferred to the Tandy in 2016. Finally, 175 ^{41}Ca samples were analyzed for two different projects covering human bioassay and environmental monitoring applications.

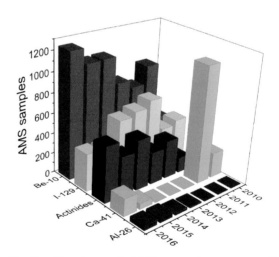

Fig. 3: *Number of AMS samples per nuclide measured since 2010.*

On the technical side, besides regular maintenance the bearings of the generator on the terminal and the GVM were replaced in 2016. Furthermore, new small Faraday Cups were installed to improve the AMS setup for ^{236}U/^{238}U.

NEW SMALL FARADAY CUPS FOR THE TANDY

Towards quasi-simultaneous measurements of $^{236}U/^{238}U$ and $^{233}U/^{238}U$

M. Christl, J. Thut, C. Vockenhuber, S. Maxeiner, H.-A. Synal

New Faraday Cups (FC) have been developed and installed on the high-energy side of the Tandy AMS system (Fig. 1). The new cups with an outer dimension of 40 x 14 mm^2 and an oval opening of 12 x 8 mm^2 are small enough to be placed at a horizontal distance of 14.6 mm which is equivalent to a mass difference $\Delta m = 3$ in the mass range of the actinides. The improved FCs are equipped with a bias ring that prevents secondary electrons from entering or escaping the FC. Together with the in-house designed current integrators precise current measurements in the low picoampere range are very well possible within a few milliseconds.

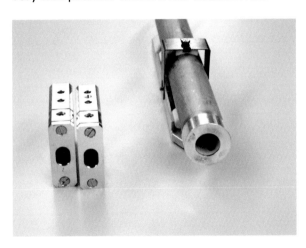

Fig. 1: *New Faraday Cups (left) compared to the previous cups (right).*

The new setup has three important new features. First, it allows measuring ^{238}U and ^{235}U currents and thus their ratio separately in two different FCs (Fig. 2). Second, with the new setup it is now possible to measure $^{236}U/^{238}U$ ratios quasi-simultaneously – very similar to the AMS setup for all other standard AMS nuclides. Such a procedure was not possible before for the actinides because of the small relative mass differences and the comparably large FCs in use. With the new FCs the Tandy, to our knowledge,

became the first AMS system that allows quasi-simultaneous $^{236}U/^{238}U$ measurements. Third, by (slowly) alternating between two settings for ^{233}U and ^{236}U (Fig. 2) both $^{233}U/^{238}U$ and $^{236}U/^{238}U$ ratios can be determined in a novel "two in one" AMS setup for uranium.

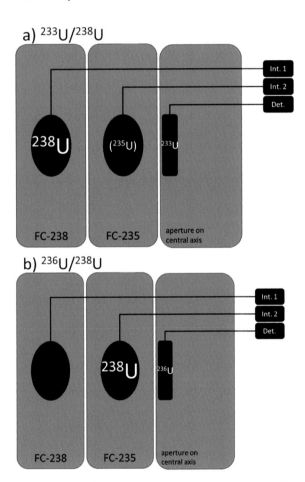

Fig. 2: *Faraday Cup and aperture setup for quasi-simultaneous $^{233}U/^{238}U$ and $^{236}U/^{238}U$ measurements (drawn to scale).*

Within the next months, the new AMS setup for ^{236}U will be implemented and the data acquisition system will be reconfigured so that routine analyses of ^{236}U will be able to take full advantage of the improved AMS setup.

A 300 KV MULTI-ISOTOPE AMS FACILITY

Latest results and current project status

S. Maxeiner, M. Christl, A.M. Müller, M. Suter, H.-A. Synal, C. Vockenhuber

A concluding proof-of-principle experiment of a novel compact multi isotope AMS facility has been conducted with the TANDY AMS. In place of the 500 kV Pelletron a MICADAS acceleration unit was used to provide (vacuum insulated) tandem acceleration, which could be upgraded to 300 kV terminal voltage [1]. It was shown before that even lower voltages around 250 kV are sufficient to measure a wide range of isotopes (actinides, ^{129}I, ^{41}Ca, ^{26}Al) efficiently and sensitively [2]. The maximum 300 kV, generated by a commercial power supply, now allowed also to investigate the feasibility of ^{10}Be AMS with such a MICADAS-type accelerator.

	[3]	Proof-of-principle		
Terminal / kV	520	300		
SiN foil / nm	75	75	75	50
Charge state	1→2	2→2	1→2	1→1
Energy / keV	740	720	430	430
Transm. / %	60	40	53	53
Std. / %	11	7	2.5	7.5
Bg / 10^{-15}	<1	1.5 ±0.9	6.3 ±2.3	19.2 ±2.2

Tab. 1: *Results of ^{10}Be AMS measurements. Charge states after stripping at the accelerator and after the SiN degrader foil are indicated.*

Results of ^{10}Be AMS measurements are shown in Tab. 1 for the TANDY AMS [3] and for three different configurations of the proof-of-principle experiment. Comparable background is only achieved with similar ion energies after acceleration, carried by the charge state 2$^+$ emerging from stripping (2nd column). The lower standard isotopic ratio of 7 % is due to the lower stripping yield of 2$^+$ (40 %) versus 1$^+$ (60 %). Ion energies of the latter are lower by 40 % and lead to increased background and decreased foil transmission due to (angular) straggling (3rd

column). A reduction in foil thickness reduces beam losses due to foil scattering (last column), but increases the background due to insufficient boron suppression. These results show the feasibility of ^{10}Be AMS on the prototype system but suggest that a more sophisticated post-acceleration spectrometer is required to increase beam transmission at low ion energies.

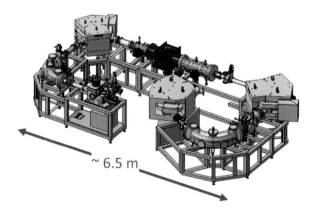

~ 6.5 m

Fig. 2: *Sketch of the novel compact multi isotope AMS system.*

Based on the experiences gained with the proof-of-principle experiments, a design for a dedicated multi isotope AMS system has been proposed (Fig. 2). An additional injection ESA provides achromatic injection to increase mass resolving power at the low energy side. Improved post-acceleration ion optics are expected with a quadrupole lens system and higher angular acceptance of magnets and ESA. The new facility is currently being developed with anticipated assembly in 2017.

[1] S. Maxeiner et al., LIP Annual Report (2015) 20
[2] S. Maxeiner et al., LIP Annual Report (2014) 13
[3] M. Christl et al., Nucl. Instr. & Meth. B 294 (2013) 29

THE ^{41}K INTERFERENCE IN ^{41}CA MEASUREMENTS

Factors related to increases in ^{41}K/^{40}Ca ratio

C. Vivo-Vilches[1], J.M. López-Gutiérrez[1], M. García-León[1], C. Vockenhuber

The main problem in ^{41}Ca AMS measurements is the interference caused by its stable isobar, ^{41}K (isotopic abundance = 6.73%). In conventional AMS facilities, this problem is solved by taking advantage of the different energy loss in the detector gas, which allows to separate the two isobars in the ΔE-E spectrum.

In compact AMS systems, such as TANDY, however, this ΔE-E discrimination becomes impossible, because the energy straggling of both isobars is higher than the difference in energy loss. Because of this, our way to estimate the ^{41}K/^{40}Ca ratio is measuring the other stable isotope of potassium, ^{39}K [1]. This K-correction has a contribution to the uncertainties directly proportional to the ^{39}K/^{40}Ca ratio.

Knowing the different ways K ions can be formed can be helpful to reduce K interference and its related uncertainty to a minimum. While extraction of (CaH$_3$)$^-$ from CaH$_2$ only leaves one way for ^{41}K interference, the production of (^{41}K^{1}H^{2}H)$^-$ anions, extraction of (CaF$_3$)$^-$ from CaF$_2$ leaves many more ways than (^{41}KF$_3$)$^-$ (M_0=98 u) due to the high mass difference of 57 u between the anion and the cation after the stripping process.

Target	^{41}M^{2+} [s^{-1}]	^{57}Fe^{2+} [s^{-1}]	^{39}K^{2+} [s^{-1}]
Al	1240	1370	16300
Blank	140	190	1400

Tab. 1: *Different average ion rates when an Al dummy or a CaF$_2$ blank sample is sputtered in the TANDY ion source.*

At TANDY, the influence of Fe presence in the cathodes has been studied because of the possibility of (^{41}K^{57}Fe)$^-$ production in the ion source [2]. Sputtering of a CaF$_2$ blank sample and an Al dummy corroborates this hypothesis.

In Tab. 1, it can be seen that Al dummies give higher ^{41}K^{2+} and ^{39}K^{2+} rates; this has been observed also in the 1 MV system at CNA in Seville: when metallic parts are sputtered, high rates from both K isotopes are measured. The HE magnet and the ESA were also tuned to detect ^{57}Fe^{2+}, selected on the LE side as (^{41}M^{57}Fe)$^-$, observing a direct relationship between rates from both K ions and ^{57}Fe^{2+} ions.

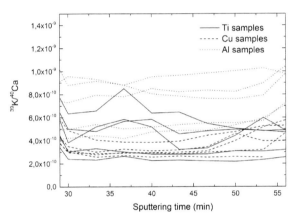

Fig. 1: *^{39}K/^{40}Ca ratio from different blank and standard samples with the same CaF$_2$:Ag mass ratio (1:9) but target holders made of different metals: titanium, copper and aluminum.*

Fe is a common trace element in most of the other metal elements. Fig. 1 shows how using Al target holders, which usually have higher Fe trace content, increases the ^{39}K/^{40}Ca ratio to about twice the one from Ti or Cu target holders.

[1] C. Vockenhuber et al., Nucl. Instr. & Meth. B 361 (2015) 273

[2] X.-L. Zhao et al., Nucl. Instr. & Meth. B 268 (2010) 816

---

[1] Centro Nacional de Aceleradores, University of Seville, Spain

CALCIUM AND STRONTIUM METABOLISM COMPARED

Study in humans using isotopic tracers and kinetic modelling techniques

J. Xiang[1], N.N. Singh[1], J.S. Sock Hoon[1], C. Vockenhuber, C. Vivo Vilches[2], H.-A. Synal, L. Lee Soon[3] and T. Walczyk[1]

Isotopic labelling of calcium in the human skeleton has been suggested as an ultra-sensitive tool to assess changes in bone calcium balance by measuring changes in tracer excretion in urine over time. A semi-stable isotope (^{41}Ca) is currently the tracer of choice for such investigations both for health as well as cost reasons. However, only a very limited number of laboratories in the world have the necessary infrastructure for ^{41}Ca detection by accelerator mass spectrometry (AMS), which has limited its wider use in research and as a potential diagnostic tool in patient care.

Here we have assessed the usefulness of a stable strontium isotope (^{84}Sr) as a surrogate tracer for ^{41}Ca for skeleton labelling. Strontium and calcium possess very similar chemical properties and are known to behave similarly in biochemical processes. In contrast to ^{41}Ca analysis, however, strontium isotopic analysis is technically less challenging using thermal ionization mass spectrometry (TIMS) or inductively coupled plasma mass spectrometry (ICP-MS) which are more widely available.

To assess the usefulness of stable strontium isotopes in studies of bone calcium metabolism, we have administered ^{84}Sr (3.6 mg) together with ^{41}Ca (3,700 Bq) as an oral bolus dose to apparently healthy, postmenopausal women (n=10) of Chinese ethnicity. Urinary excretion (24 h) of both tracers was studied over 6 months using AMS for ^{41}Ca analysis (Fig. 1) and TIMS for ^{84}Sr analysis. Kinetics of tracer excretion were assessed using a three-compartment model of human calcium metabolism.

We found that long-term kinetics of strontium metabolism and calcium metabolism are very similar and can be described using the same compartment model. As such, labelling of the

skeleton with strontium isotope tracers permits the detection of changes in bone mineral balance, in principle, as for the ^{41}Ca technique. However, differences in renal clearance between both elements must be taken into consideration for correct interpretation of findings.

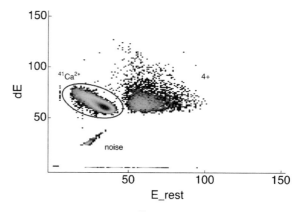

Fig. 1: Detection of ^{41}Ca at the 500 kV TANDY compact AMS facility. ^{41}Ca^{2+} ions are identified in the gas ionization detector by its energy loss signals in the first anode (dE) and second anode (E_rest).

[1] NutriTrace@NUS, National University of Singapore
[2] Centro Nacional de Aceleradores, Sevilla, Spain
[3] Department of Medicine, National University of Singapore

THE MICADAS AMS FACILITY

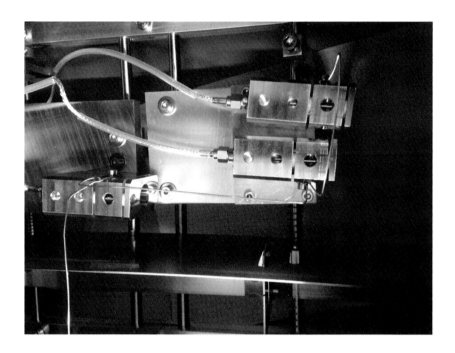

Radiocarbon measurements on MICADAS in 2016

Installation of GeoLipMICADAS

RADIOCARBON MEASUREMENTS ON MICADAS IN 2016

Performance and sample statistics

Scientific and technical staff, Laboratory of Ion Beam Physics

In the past years the ProtoMICADAS, which has been operating for more than 10 years, came to its limits with more than 10000 analysed samples per year.

Thus a new GeoLipMICADAS was installed in 2015 and taken into operation in January 2016. Already more than 5500 samples were measured primarily as graphite in the first year of operation on the new GeoLipMICADAS! Most gas samples were still measured on the ProtoMICADAS (8200). Over all, more than 13700 samples were analysed in 2016 (*Fig. 1 and Fig. 2*), an increase of 35% over 2015! We still see a continuous trend towards gas measurements with an increase of nearly 50%. Nevertheless, also the amount of graphite samples increased by 20%.

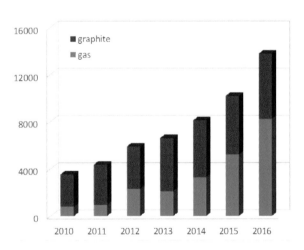

Fig. 1: The amount of radiocarbon samples analysed steadily increase over the years. Nearly 14000 sample targets were measured in 2016.

The two instruments allowed us to analyse graphite samples for longer. Additionally, the GeoLipMICADAS will be soon equipped with improved current integrators that will allow us to measure currents up to 180 µA C⁻ (previously 90 µA C⁻). Routine measurements now run typically with ion currents of 90 to 120 µA C⁻

from the ion source and high-energy current of 40 to 60 µA C⁺. The longer measurement times combined with higher currents allow us to measure routinely with higher precision.

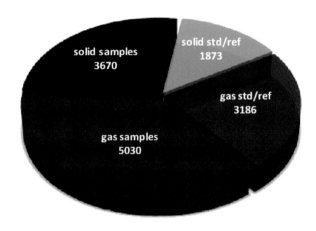

Fig. 2: Samples measured on MICADAS in 2016. Graphite samples and standards are in blue. Red indicates samples measured with the gas ion source.

About 20% more samples (3300) than in previous years were measured for our partner institution. The amount of commercially measured samples increased by 25% (2500), namely because significantly more gas samples were measured (small-size carbonate samples). The internally measured samples were also increased to 2000 from 1500 in the previous year. A major contribution comes here from a research project that aims to extend the tree-ring based calibration curve [1].

[1] A. Sookdeo et al., Annual report 2016, 31

Installation of GeoLipMICADAS

Pushing the limits of highest precision ^{14}C measurements

L. Wacker, H.-A. Synal

The GeoLipMICADAS was taken into operation in January 2016 and has, since February 2016, already been used for routine and highest precision measurements of samples converted to graphite. By the end of 2016, the GeoLipMICADAS was additionally used for gas measurements (Fig. 1).

The GeoLipMICADAS uses helium as stripper gas and thus clearly exceeds the performance of the ProtoMICADAS. However, several modifications were done in the first year of operation to improve its performance. While a newly installed source extraction lens did not significantly improve the performance, reducing the aperture size from 12 mm to 8 mm after the ESA in front of the detector improved the background for processed blanks from approx. 48 000 years to about 52 000 years.

Fig. 1: *The new GeoLipMICADAS with a new gas interface from Ionplus at ETHZ.*

In order to receive higher measurement precision in shorter times, the current

integrators were modified to accept ^{12}C$^+$ ion currents of up to 80 µA instead of 40 µA and ^{12}C$^-$ currents of 200 µA instead of 100 µA can now be measured. Routine measurements run now between 80 and 120 µA ^{12}C$^+$ currents. A set of 39 samples and standards can now be measured for precisions of 2‰ within one day.

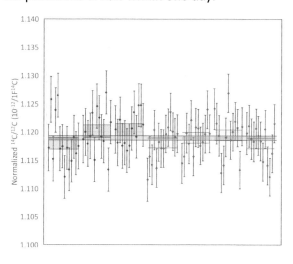

Fig. 2: *The figure shows the normalized ratios measured for all seven standards in a high-precision measurement. The OX-1 (orange) and OX-2 are in good agreement. The given 1-σ-errors are purely based on counting statistics.*

An example of seven standards for highest-precision within 42 hours is given in Fig. 2. Three OX-1 and four OX-2 standards match within counting statistic errors of 1‰ or less. The data were acquired within less than two days at negative ion currents close to 100 µA and transmissions of over 48%! Absolutely measured ^{13}C/^{12}C ratios are only 1-2% lower than their nominal value, while ^{14}C/^{12}C values are about 5% lower. Overall, the new GeoLipMICADAS has a 30% higher ^{14}C yield than the ProtoMICADAS with a transmission of only 40% and 15% lower ^{14}C/^{12}C ratios.

DETECTION AND ANALYSIS

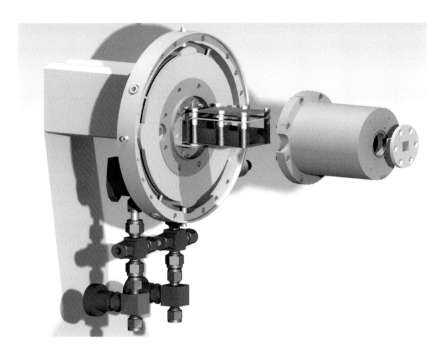

Ultimate energy resolution at lowest energies

Beam Profile Monitor

ULTIMATE ENERGY RESOLUTION AT LOWEST ENERGIES

Operating the ETH gas ionization chamber in proportional mode

A.M. Müller, M. Döbeli, H.-A. Synal

The performance of gas ionization chambers (*GIC*) at very low energies (< 100 keV) is essentially determined by the electronic noise level of the charge sensing preamplifier. Optimizing the detector signal height per energy loss would reduce the impact of the electronics on the detector resolution. One approach of increasing the detector signal is to operate the GIC in the so-called proportional detection mode. At high electric field strengths the electron signal is amplified by secondary ionization processes of the detector gas, while the resulting signal amplitude remains proportional to the deposited energy.

We made experiments with the ETH GIC [1], where secondary ionization was initiated between the Frisch-grid and the charge collecting anode. Fig. 1 shows the detector pulse height and the relative resolution for protons at 300 keV as a function of the reduced electric field strength *E/p* (*p* being the gas pressure) between grid and anode.

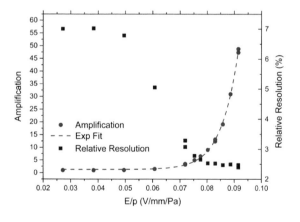

Fig. 1: *Electron amplification factor and relative energy resolution as a function of the reduced electric field strength for protons at 300 keV [2].*

Initially, at low field strength, the detector is operating as a conventional GIC, where primary ionization charge carriers are collected. At higher values of *E/p* secondary ionization processes appear and the detector signal height grows exponentially. At the same time the relative energy resolution drops due to a decreasing significance of electronic noise and approaches the physical limit determined by the variation of the charge carrier production process and the energy straggling in the detector entrance window.

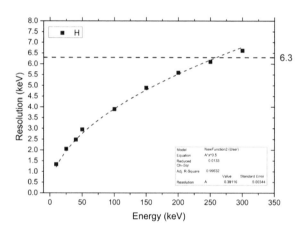

Fig. 2: *Energy resolution for protons for the GIC operated in proportional mode. The dashed line at 6.3 keV corresponds to the level of electronic noise obtained with the conventionally operated GIC equipped with Peltier cooled preamplifiers.*

In this way it is possible to overcome the limitations due to electronic noise of a conventionally operated GIC and even ions at energies of a few tens of keV can be detect and resolved (Fig. 2).

[1] A.M. Müller et al., Nucl. Instr. & Meth. B 287 (2012) 94

[2] A.M. Müller et al., submitted to Nucl. Instr. & Meth. B

BEAM PROFILE MONITOR

Device design and phase space measurements

D. De Maria, H.A. Synal, A.M. Müller, S. Maxeiner

The phase space occupied by the negative ion beam from the source of an AMS system, given by its width and angular divergence, is a key performance figure as it strongly influences the quality of beam transport through the system.

To investigate the phase space generated by a MICADAS type ion source, a Beam Profile Monitor (BPM) based on two oscillating sensing wires located at two different positions along the beam axis was developed (Fig. 1). The versatile measurement board Red Pitaya, implemented in the experimental set-up, has been programmed to provide the driving signal for the wires taking the intensity profile of the beam, as well as data acquisition of the measured beam current. Algorithms to determine parameters such as beam width and divergence were developed. In addition, the relative position of the BPM to a reference beam was calibrated using analytical formulas.

Fig. 1: *Picture of the developed BPM. The instrument is mounted on the MyCadas facility after the source dipole magnet.*

Measurements of the behaviour of the phase space have been performed as a function of the beam energy in the range between 30 keV and 50 keV at the MyCadas facility. The measurements taken after the source magnet allowed the identification of a phase space at

FWHM of (4.54 ± 0.28) mm·mrad·$\sqrt{\text{MeV}}$ in the horizontal (dispersive) plane (Fig. 2) and of (3.42 ± 0.21) mm·mrad·$\sqrt{\text{MeV}}$ in the vertical plane (Fig. 3). The latter is not influenced by the beam's energy spread and therefore it corresponds to the effective beam phase space.

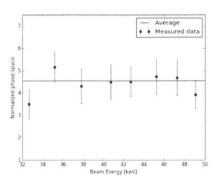

Fig. 2: *Phase space in the horizontal plane normalized with the beam energy.*

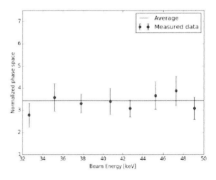

Fig. 3: *Phase space in the vertical plane normalized with the beam energy.*

Due to the ion source geometry, the formed beam is expected to be circular symmetric. Thus, a relative beam energy spread of the order of 10^{-3} can be estimated from the difference in phase space by taking into account the energy dispersion of the magnet in the horizontal direction.

RADIOCARBON

Terrestrial cycling of organic matter

Revisiting millennial-scale climate variability

Organic carbon export during extreme events

^{14}C SAMPLES AT ETH LABORATORY IN 2016

Overview of samples prepared for ^{14}C analysis

I. Hajdas, L. Hendriks, M. Maurer, M. B. Röttig, A. Sookdeo, H.-A. Synal, L. Wacker, C. Welte, C. Yeman

Accurate radiocarbon chronologies require selection of proper carbon fraction and/or chemical treatment, which removes potential contamination. Figure 1 illustrates the great variety of applications that apply the ^{14}C dating method. This is also reflected in a wide range of material submitted to the ETH laboratory (Fig. 2).

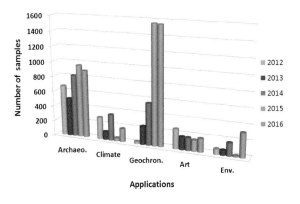

Fig. 1: *Number of samples (objects) analysed for various research disciplines during the last five years.*

Our ongoing research projects have resulted in higher throughput of samples in 2016 (ca. 400 more than 2015). Nearly 70% of the samples in 'Geochronology' is wood, which was analysed as a part of internal projects.

Research	Total	Internal
Archaeology	897	26
Past Climate	180	12
Geochronology	1553	1057
Art	193	35
Environment	322	185
Total	**3145**	**1315**

Tab. 1: *Number of samples analyzed in 2016 for various applications. Column 'Internal' is the number of samples supported by the laboratory for master or term theses.*

Environmental studies form a considerable share of internal projects. An increasing number of art related samples was collected this year as a part of a LIP project.

Multiple AMS analyses are required for samples such as mortar or canvas with paint, where more than one fraction is collected. Often, analyses of contaminates are a source of information about the history of objects.

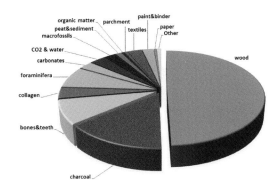

Fig. 2: *Type of material prepared and analysed at the ETH laboratory in 2016.*

In summary, during 2016 numerous interesting questions could be answered with the help of ^{14}C dating including the mummified remains of Queen Nefertari [1]

[1] M.E. Habicht et al., Plos One 11 (2016) 1

RADIOCARBON DATING OF PAINTINGS

^{14}C analysis of binding medium for artwork dating

L. Hendriks, I. Hajdas, M. Küffner[1], N. Scherrer[2], S. Zumbühl[2], E. Ferreira[3]

A painting consists primarily of a support material (wood, paper, canvas), which is coated with a ground layer on top of which pigments mixed in an organic binder are applied. Nowadays when the authenticity of a work of art is questioned and ^{14}C measurements are requested, only the support material is dated. However, the results may not always be conclusive as a possible later use of an older artwork's support cannot be excluded. The natural origin of the binder makes it an ideal material for ^{14}C dating (Fig. 1). Additionally it cannot be re-used and hence it is representative of the time of creation.

Fig. 1: *"Bildnis Margrit mit roter Jacke und Konzertkleid" painted in 1962 by Franz Rederer (1899–1965). ^{14}C ages of binding medium with 'bomb peak' predate the painting by 4-5 years.*

The challenge remains in finding a suitable sampling location, as was demonstrated by Hendriks et al [1]. Indeed, prior to ^{14}C analyses, a detailed characterization of all paint components in the sampled material is crucial. This is achieved by combining XRF, FTIR and Raman spectroscopy to identify the binder type, filler and pigments (Fig. 2). Ideally, inorganic pigments are the paint samples of choice as they are not primarily carbon based. White

paint, for instance, is often made of lead white, zinc white, titanium white or a mixture thereof and hence is an excellent candidate for further ^{14}C analysis.

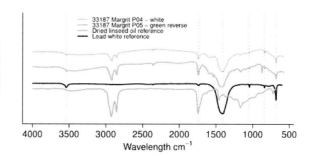

Fig. 2: *FTIR spectra of paint samples (grey and green) in comparison to reference spectra of lead white (black) and linseed oil (yellow).*

The presence of carbonates in the lead white pigment, at first thought to be problematic, was resolved by treating the sample with acid in excess, hereby forming a salt, water and carbon dioxide.

$$PbCO_3 + HCl \rightarrow PbCl_2 + H_2O + CO_2$$

The white paint, carbonate free and weighing only a few hundred micrograms, was successfully measured on the gas ion source of the MICADAS AMS. Resulting ^{14}C ages on the binding medium predate the painting by 4-5 years and correlate with the ^{14}C ages of the canvas. These results illustrate the potential of dating the binding medium, which offers a new tool in the study of paintings and their origin.

[1] L. Hendriks et al, Appl. Phys. A 122 (2016)

[1] *SIK-ISEA, Swiss Institute for Art Research, Zürich*
[2] *HKB, Bern University of Arts*
[3] *TH Köln, University of Applied Sciences, Germany*

THE 775 AD COSMIC EVENT

World-wide measurements of the 775 AD event in tree-rings

L. Wacker, D. Galvan[1], U. Büntgen[1], J. Wunder[1], H.-A. Synal

Annual resolution and absolute dating are essential for high-resolution proxy archives, as well as for the accurate dating of archaeological and historical evidence. Although cosmic events can generate distinct atmospheric radiocarbon peaks, their detection is still restricted to a few records and sites.

Here, we measured the ^{14}C content of 385 disjunct tree-ring samples between 770 and 780 AD from different dendrochronological records around the word (Fig. 1) within a joint project, called COSMIC, between WSL Birmensdorf and ETH Zurich. The period of time was selected, because a strong solar proton event [2] in 774/775 AD likely triggered the global extent of a radiometric time marker.

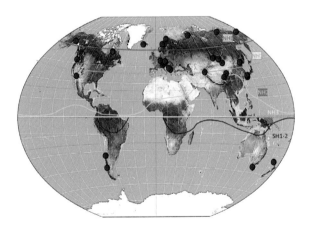

Fig. 1: *Locations of the sampled dendro records measured for the COSMIC project. While most records are located on the Northern Temperate and Arctic Zone (24), also seven records from the Southern Temperate zone were analysed.*

A compilation of the measured ^{14}C signals is given in Fig. 2. Significant ^{14}C spikes in 775 AD [1] are synchronous between 25 tree species at 31 sites on five continents and both hemispheres. The measurements demonstrate that the tested dendrochnological records are robust back to 774 AD with no counting errors.

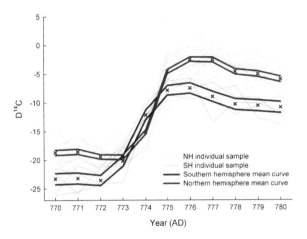

Fig. 2: *Measured radiocarbon concentrations (as Δ^{14}C) for all measured records (fine line) with it's mean concentrations (crosses with solid line) for the Northern Hemisphere (blue) and the Southern Hemisphere (red).*

Given the cross-disciplinary importance of fast cosmic events, we advocate reproducing a yearly-resolved IntCal calibration curve.

[1]   F. Miyake F., Nature 486 (2012) 240

[2]   I.G. Usoskin, A&A L3 (2013) 552

[1] WSL, Birmensdorf

PREHISTORIC COPPER MINING IN THE GRISONS ALPS?

^{14}C dating and dendrochronology for absolute dating

R. Turck[1], M. Oberhänsli[2], M. Seifert[2], A. Sindelar[1], I. Hajdas, L.Reitmaier-Naef[1]

Excavations and investigations of prehistoric copper mines located in the southeastern part of Switzerland, canton of Grisons (GR) in the Oberhalbstein Valley are performed in a collaboration between the Universities of Innsbruck (A) and Zurich (CH), the Deutsches Bergbaumuseum Bochum (D) and the Curt-Engelhorn-Centre for Archaeometry in Mannheim (D).

Fig. 1: Smelting furnace, Marmorera, "Gruba".

One of the main questions concerns the age of one known mining gallery, numerous slag deposits and smelting sites including archaeological features like smelting furnaces (Fig 1). When excavating and prospecting large scale areas large amounts of wood charcoal were found at most sites (nearly 70 known) for smelting and mining [1]. Mostly larch, spruce and pine were used for firing the ovens or to set fire in the mining gallery.

Well preserved charcoal sometimes having more than 150 year rings is excellent for dendrochronology (Fig. 2). The trees at high elevations of 1000 to 2300 m a.s.l. grow very slowly and have close annual rings. So far, 48 samples could be synchronized and transferred to the standard sequences of the Central and Eastern Alps. This dating leads us to the 8th and 7th century BC: The Early Iron Age.

In few cases ^{14}C dating was applied for cross-checking of dendrochronology. Moreover, 25 ^{14}C samples from prospected sites (mostly smelting sites) with poor tree ring preservation were analyzed. Some fragments show the „old wood effect" i.e., are older than ages obtained by ^{14}C and dendrochronology.

Future ^{14}C and dendro analysis of Oberhalbstein charcoal will focus on the question whether bronze copper mining and smelting can be proved in the Oberhalbstein.

Fig. 2: Charcoal dating to 685 BC, having 161 tree rings.

[1] P. Della Casa et al., Quat. Int. 402 (2016) 26-34

[1] Insitute for Archaeology, University of Zurich
[2] Archaeological Service of Grisons, Chur

FILLING A VOID

^{14}C DATES FOR LATE GLAICAL TREES BETWEEN 12.3 – 11.9 ky BP

A. Sookdeo, L. Wacker, M. Friedrich[1,2], F. Reinig[3], B. Kromer[2], U. Büntgen[3], M. Paulyl[4], G.Helle[4], H.-A. Synal

The onset of the Younger Dyras (YD) around 12900 BP is marked by an abrupt cooling event lasting over 1000 years before a warming to present day temperatures. The abrupt climate change during the YD could be a potential analog to modern day global climate change. To understand the YD, accurate and complete paleoclimate, paleoenvironmental and archaeological records are required. However, Northern Hemisphere absolute tree-ring chronologies extend back to only 12325 years BP and floating chronologies exist back to only to 14200 years BP. Radiocarbon dates for the absolute tree-chronologies are decadal averages and are weakened as there are only 12 decadal dates for between 12325 and 11900 years BP [1]. At ETH-Zurich we have measured quinquennial radiocarbon between 12325 and 11900 years BP using german trees that are part of the absolute tree-ring chronology (Fig. 1).

Following the guidelines of the International Calibration (IntCal) the samples were also prepared and measured by another laboratory, in this case the University of Mannheim (Fig. 1). Visual comparison of the two datasets shows no statistical difference; errors are shown within 1σ.

This data plotted along with Intcal13, adds more structure to the radiocarbon record and successfully fills the void between between 12.3-11.9kBP (Fig. 2).

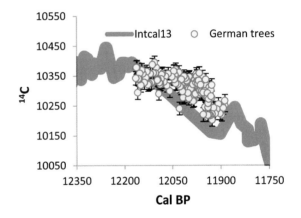

Fig. 2: *Ages of german trees plotted along with Intcal13 (see text).*

[1]   A. Hogg et al. Radiocarbon 58 (2016) 947

[2]   P. Reimer et al. Radiocarbon 55 (2013) 1869

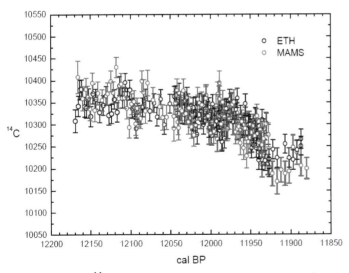

Fig. 1: *^{14}C dates for German trees measured at ETH and Mannheim (MAMS).*

[1] *Environmental Physics, Heidelberg University, Germany*
[2] *Botany, Hohenheim University, Stuttgart, Germany*
[3] *Swiss Federal Research Institute, WSL, Birmensdorf*
[4] *Helmholtz Centre Potsdam, GFZ, Germany*

THE REVISED SETUP OF THE LASER ABLATION ^{14}C-AMS

Reaching higher ion currents

C. Yeman, C. Welte, B. Hattendorf[1], J. Koch[1], M. Christl, L. Wacker, H.-A. Synal

The fastest way to analyze carbonates for radiocarbon is using the novel LA (Laser Ablation)-AMS technique [1]. By focusing a pulsed laser beam of UV light on the sample surface, CO_2 is produced, which is directly and continuously introduced into the gas ion source of the MICADAS and analyzed online for radiocarbon. A positioning system allows precise movement of the sample relative to the laser beam. Hence, scanning along the growth axis of a sample allows recording a continuous ^{14}C profile.

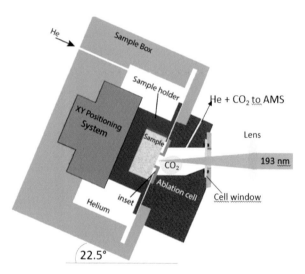

Fig. 1: *Schematic view of the improved setup of the laser ablation cell.*

The setup was revised targeting the limiting factors of the CO_2 production in the ablation unit and therefore the low ion currents at the AMS. Primarily, this was owed to low laser fluences on the sample which were restricted by the damage threshold of the cell window. The fluence – the energy of the laser beam per area - is driving the ablation process and determines the CO_2 conversion efficiency. Thus, a new cell head was designed which is shown in Fig.1. The distance between the cell window and the sample was doubled and a new lens with

shorter focal length is now used to focus the laser onto the sample. Both changes reduce the fluence on the cell window, increasing its lifetime and no longer limiting the fluence on the sample. At the same time, a smaller spot size is achieved with the new lens, leading to higher fluences on the sample and higher spatial resolution down to 75 µm. In addition, the optical setup has been improved by halving the optical length and minimizing the number of mirrors to reduce energy losses of the laser. In Tab. 1 the key characteristics of the original and the revised setup are listed. The maximum achievable fluence on the sample is now 23 J/cm^2 compared to 2.5 J/cm^2. With the new setup ion currents at the MICADAS have doubled.

	Original Setup	Improved Setup
Distance window-sample	12.5 mm	25 mm
Focal length lens	5.5 cm	4 cm
Volume	0.6 ml	0.9 ml
Energy on sample	0.8-1 mJ	0.8-2.4 mJ
Spot size	680x110µm^2	140x75µm^2
Fluence on sample	1-2.5 J/cm^2	8-23 J/cm^2

Tab. 1: *Comparison of the key characteristics of the original and the improved setup.*

[1]   Welte et al., Anal. Chem. 88 (2016) 8570

[1] *DCHAB, ETH Zurich*

HOW LONG CAN RED SNAPPER LIVE?

Revealing the bomb radiocarbon pulse in an otolith with LA-AMS

C. Yeman, A.H. Andrews[1], C. Welte, B. Hattendorf[2], L. Wacker, M. Christl, H.-A. Synal

The otolith of a red snapper (*Lutjanus campechanus*) was analyzed for radiocarbon (^{14}C) with a novel laser ablation (LA) apparatus coupled to the MIni CArbon DAting System (MICADAS). By prior growth zone counting in the otolith it was estimated that the fish was 50-55 years old. The cross section of the otolith (Fig. 1) used in the LA scans was only a few hundred micrometers thick and scanning distances did not exceed 6.5 mm. In this application, the LA-AMS technique excels due to low material usage, high spatial resolution and a quasi-continuous ^{14}C signal along the growth axis.

Fig. 1: *Cross section of the red snapper otolith: on the left side material was ablated by the laser in three different positions and on the right side samples were extracted by micromilling for individual CO_2-AMS measurements.*

Material was ablated in several positions of the otolith section—starting close to the core (earliest growth zones), moving along the growth axis to the outer edge, and then returning to the core in the opposite direction along the same path. In addition, a series of samples were taken along the growth axis by micromilling and the extractions were analyzed individually by CO_2 measurement on the MICADAS—the aim was to corroborate the findings from LA-AMS. These data were compared to a bomb ^{14}C reference record derived from corals of the Gulf of Mexico [1], where this fish species lives and was collected for this study. All laser scans and micromilled samples revealed the bomb ^{14}C peak and pre-bomb levels. The rise of bomb ^{14}C was located on the otolith. Growth zone counting added ~12 years prior to the ^{14}C rise. The coral ^{14}C record indicated the rise was in 1958; the red snapper captured alive in 2004 was ~58 years old.

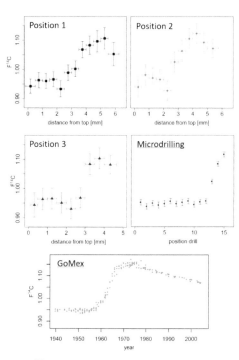

Fig. 2: *$F^{14}C$ Data for the LA-AMS scans at positions 1-3 (Fig. 1), micromilling series, and the GoMex coral ^{14}C reference record.*

Analysis of the otolith by LA-AMS validated the age estimation procedures for red snapper and increased longevity to near 60 years.

[1] Andrews et al., Can. J. Fish. Aquat. Sci. 70 (2013) 1131

---

[1] *NOAA Fisheries, Pacific Islands Fisheries Science Center, Hawaii USA*
[2] *DCHAB, ETH Zurich*

SEDIMENT STORAGE ON THE BOLIVIAN ALTIPLANO

Dating fluvial terraces with radiocarbon

T. Gordijn[1], K. Hippe, I. Hajdas, S. Ivy-Ochs, V. Picotti[1], M. Christl

Sediment, on its way from source to sink, can be temporarily stored in a fluvial catchment. In this study, we investigate the nature, age and spatial distribution of sediment storage in a catchment on the eastern border of the Bolivian Altiplano [1]. The duration of sediment storage in fluvial terraces and alluvial fans provides information on the climatic and tectonic situation in the study area and their changes during the late Quaternary.

Following detailed mapping of storage deposits in the field, organic-rich sediment layers from two different fluvial terraces were sampled for radiocarbon analysis (Fig. 1). The sediment was sieved and the macro remains of plants >250 μm were handpicked for dating. Depending on the weight of the sample it was either graphitized or measured using GIS. A bone fragment found in the basal gravels of another higher fluvial terrace was also dated.

Fig. 1: *a) Organic-rich layer in a fluvial terrace. b) Microscope picture of organic-rich sediment. c) Separated organic particles >250 μm, red arrows indicate younger rootlets and contaminants. d) Analyzed bone fragment.*

The three terrace levels identified in the study area and their corresponding radiocarbon ages for the higher and lower terraces are shown in Fig. 2. The two radiocarbon profiles are from the lower terrace level and both profiles get consistently younger from bottom to top. The profile situated in a tributary stream yields younger ages (by roughly 1000 years) compared to the profile dated from the main valley. This can be explained by the fact that the main valley acts as the base level for the tributary streams.

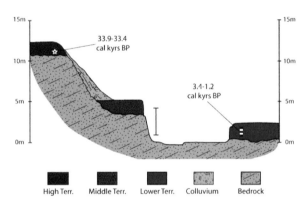

Fig. 2: *Schematic profile of the three terrace levels and their calibrated radiocarbon ages.*

Results indicate that both, terrace deposition and incision, take place during high precipitation sequences of a climate cycle. Converting our ages into rates of channel incision, the uplift in the area seems to be higher than previously expected but agrees with the proposed crustal rebound caused by focused erosion of the Rio La Paz [2].

[1] K. Hippe et al., Geomorph. 179 (2012) 58
[2] G. Zeilinger and F. Schlunegger, Terra Nova 19 (2007) 373

[1] *Geology, ETH Zurich*

STALAGMITE RADIOCARBON CHRONOLOGIES

A novel approach for the construction of reliable ^{14}C chronologies

F.A. Lechleitner[1], J. Fohlmeister[2], C. McIntyre[1], L.M. Baldini[3], R. A. Jamieson[3], H. Hercman[4], M. Gasiorowski[4], J. Pawlak[4], K. Stefaniak[5], P. Socha[5], T.I. Eglinton[1], J.U.L. Baldini[3]

Stalagmites are important paleoclimate archives, and can often be dated with exceptional precision using U-series methods. However, a number of stalagmites cannot be dated using U-series, and for those samples ^{14}C could be an alternative. ^{14}C-dating of stalagmites is complicated because of the variable reservoir effect (dead carbon fraction, DCF) introduced by the dissolution of radiocarbon-dead host rock [1, 2].

We developed a method that allows the construction of robust chronologies for stalagmites using ^{14}C [3]. The technique calculates a best fit between a series of ^{14}C measurements on a stalagmite and known atmospheric ^{14}C activity using the radioactive decay equation (Fig. 1).

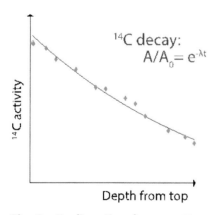

$$^{14}C \text{ decay:} \quad A/A_0 = e^{-\lambda t}$$

Fig. 1: Radioactive decay pattern (black line) in measured stalagmite ^{14}C data (green diamonds).

With one anchor point of known age (e.g., the ^{14}C bomb spike, or one reliable U-Th date), the model accounts for DCF. The model was tested on a stalagmite from Heshang Cave, China (Fig. 2), and on a previously un-dateable stalagmite from Niedźwiedzia Cave, southern Poland, achieving very good results.

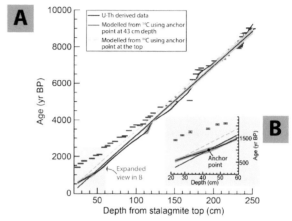

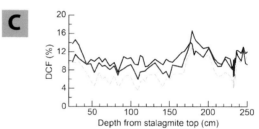

Fig. 2: Results from age modeling in stalagmite HS4 from Heshang Cave, China (further details in [3]).

[1] D. Genty et al., GCA 65 (2001) 3443
[2] M.L. Griffiths et al., Quat. Geochron. 14 (2012) 81
[3] F.A. Lechleitner et al., Quat. Geochron. 35 (2016) 54

[1] Geology, ETH Zurich
[2] Institute of Environmental Physics, University of Heidelberg, Germany
[3] Department of Earth Sciences, Durham University, UK
[4] Institute of Geological Sciences, Polish Academy of Sciences, Poland
[5] Institute of Paleozoology, University of Wroclaw, Poland

ORGANIC CARBON IN DANUBE RIVER SEDIMENTS

Ageing of organic carbon during transport along a major river

C.V. Freymond[1], F. Peterse[2], F. Filip[3], L. Giosan[4], N.Haghipour, L.Wacker, T.I. Eglinton[1]

Through erosion processes, soil material, plant litter and rock debris is mobilized and transported within river basins. Rivers are the main connection between the terrigenous and the oceanic carbon pool, exporting about 200 Mt/yr biospheric and petrogenic carbon to the world oceans [1]. Deposited on continental margins, the sediment including organic biomarkers builds up an important archive of environmental conditions on the continent, e.g. temperature, vegetation type. To interpret these sediment deposits, it is important to better understand the transport history of organic carbon from the source of production to the sedimentary sink.

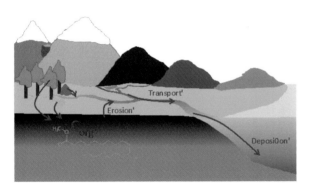

Fig. 1: *Simplified schematic of organic carbon transport in a river system from source to sink.*

In this project, recent river sediment from the Danube River (Fig. 2), the largest river in the European Union is investigated.

^{14}C measurements of the bulk sediment and the fine fraction (i.e. <63 μm grain size) give an overall picture of the ^{14}C activity in the sediment, including biosperic and petrogenic carbon. To get a better picture of plant derived carbon, compound specific ^{14}C measurements of plant wax fatty acids were done.

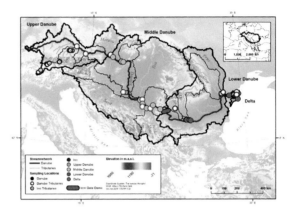

Fig. 2: *The Danube River basin including sampling locations.*

Here we show preliminary ^{14}C data of bulk sediment and the fine fraction. Along the Danube River, a strong trend to increasing ages is visible.

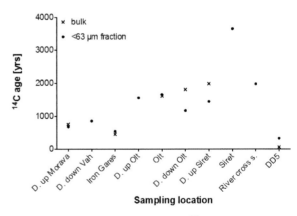

Fig. 3: *Bulk and fine fraction ^{14}C age along the Danube River.*

[1] V. Galy et al., Nature, 521 (2015) 204

[1] *Geology, ETH Zurich*
[2] *Geochemistry, Utrecht University, The Netherlands*
[3] *Department of Geography, University of Bucharest, Romania*
[4] *Woods Hole Oceanographic Institution, USA*

TRACING PYROGENIC CARBON ON A GLOBAL SCALE

Elucidating the turnover and C sequestration potential of pyrogenic C

D.B. Wiedemeier[1], A.I. Coppola[1], T.I. Eglinton[2], N. Haghipour[2], Cameron McIntyre[2], M.W.I. Schmidt[1]

Combustion-derived, pyrogenic carbon (PyC), is the organic carbon fraction that includes charcoal and other burned organic residues. It is relatively resistant against chemical and biological degradation in the environment, possibly leading to a slow turnover, which supports carbon sequestration. PyC is produced on large scales (hundreds of teragrams) in biomass burning events such as wildfires, and by combustion of fossil fuel in industry and traffic.

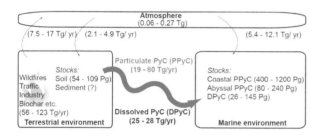

Fig. 1: *A rough estimate of the global PyC cycle (modified from Bird et al. 2015) [1].*

PyC is an inherently terrestrial product and thus has predominantly been investigated in soils and the atmosphere. Much fewer studies are available about the subsequent transport of PyC to rivers and oceans. Recently, awareness has been rising about the mobility of PyC from terrestrial to marine systems and its fate in coastal and abyssal sediments was recognized. It is therefore crucial to extend our knowledge about the PyC cycle by tracing PyC through all environmental compartments. By comparing its biogeochemical behavior and budgets to that of other forms of organic carbon, it will be possible to elucidate PyC's total spatiotemporal contribution to carbon sequestration.

In this study, we use a state-of-the-art PyC molecular marker method [2] to trace quantity, quality as well as ^{13}C and ^{14}C signature of particulate PyC (PPyC) in selected major river

systems around the globe. Previous studies suggested a distinct relationship between the ^{14}C age of plant-derived suspended carbon and the latitude of the river system, indicating slower turnover of plant biomarkers in higher latitudes (cf. Fig 2). Our preliminary results show that PPyC shows a similar latitudinal trend. Moreover, PyC seems indeed older than other forms of terrestrial carbon, implying a slower turnover than other forms of organic C and therefore a high potential for carbon sequestration on a global scale.

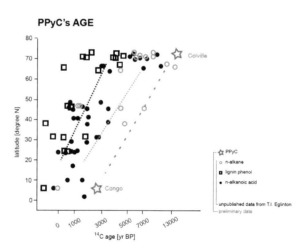

Fig. 2: *Age as a proxy for the turnover of PPyC and other forms of organic carbon in major river systems (preliminary PPyC results, terrestrial biomarker data from T.I. Eglinton, unpublished).*

[1] M.I. Bird et al., Annu. Rev. Earth Planet Sci. 43 (2015) 273

[2] D. B. Wiedemeier et al., JoVE 111 (2015) 1

[1] *Geography, University of Zurich*
[2] *Geology, ETH Zurich*

TIMESCALES OF CARBON EXPORT TO THE BENGAL FAN

Modeling the bomb spike recorded in terrestrial fatty acids

K. French[1], C. Hein[2], L. Wacker, N. Haghipour[3], T. Eglinton[3], V. Galy[1]

Estimating the timescales of terrestrial organic carbon transport to marine sediments is key to integrating the terrestrial carbon cycle into the global carbon cycle and essential to interpreting paleo-records. The Ganges-Brahmaputra system is an ideal locality to study the timescales of continental storage and riverine transport of terrestrial organic matter to marine sediments. Here, we use bomb carbon as a tracer in order to estimate turnover timescales. Fatty acids and bulk organic matter from a Bengal Fan sediment core record the bomb spike, but the magnitude is smaller than the perturbation recorded in the atmosphere. Clearly, a fast-cycling component of terrestrial carbon is incorporating bomb carbon while a slower cycling component void of bomb carbon is diluting the signal. Modeling reveals that a mixture of a decadal component with an average age of 14 to 24 years and millennial component with an average age from 900 to 1600 years best approximates the fatty acid ^{14}C results (Figs. 1 and 2). Millennial carbon contribution (fMill) increases with chain length.

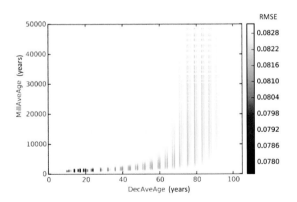

Fig. 1: *The heat map of root mean square error (RMSE) shows the quality of fit between synthetic and measured long chain fatty acid data for ~60,000 combinations of millennial and decadal average ages. The best fitting combinations are colored in dark red.*

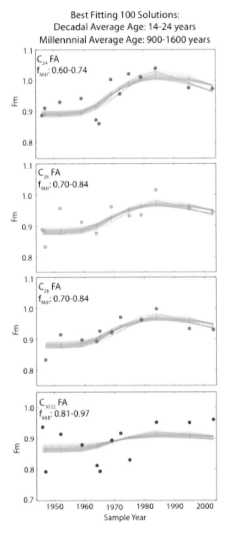

Fig. 2: *The synthetic model results from the top 100 combinations of decadal and millennial components are plotted (grey) in comparison to the measured fatty acid (FA) radiocarbon data.*

[1] WHOI, Woods Hole, USA
[2] Virginia Institute of Marine Science, USA
[3] Geology, ETH Zurich

ROLLING IN THE DEEP

Temporal constraints on tectonically-triggered events in the hadal zone

R. Bao[1,2], M. Strasser[1,3,4], A. McNichol[2], N. Haghipour[1], C. McIntyre[1], G. Wefer[4], T. Eglinton[1]

The origin, nature and past variability of sediments accumulating in the abyssal ocean is a topic that has garnered the attention of many geoscientists. Sediment deposits in deep ocean trenches, one important type of hadal environment located in tectonically-active regions, hold great potential for understanding large-scale sediment remobilization and translocation processes triggered by major earthquakes, and for documenting the past history and frequency of such events. Establishing the chronostratigraphic framework for hadal zone sedimentary records constitutes a long-standing issue as they are deposited below the Calcite Compensation Depth (CCD), resulting in an absence of dateable (i.e. carbonate biominerals), thereby confounding traditional ^{14}C dating methods. This is one of the most critical challenges that must be overcome in order to constrain the provenance and frequency of specific event deposits, and to link them to specific earthquakes.

In this study, we present results from detailed radiocarbon-based investigation of the organic matter in a sediment core retrieved from the Japan Trench (> 7.5 km water depth), proximal to the giant Tohoku-oki earthquake and ensuing tsunami of 2011. Construction of a high temporal resolution bulk organic carbon (OC) ^{14}C record (Fig. 1) reveals that sedimentation in the Japan Trench is interrupted by episodic deposition of sediments characterized by pre-aged OC. These sedimentary layers coincide with intervals that have been attributed to past, historically-recorded earthquakes [1]. Moreover, we describe further ^{14}C measurements on specific thermally-resolved organic matter fractions from ramped pyrolysis-oxidation of a subset of sediment samples that yield new chronological constraints in the context of past earthquake history in the Japan Trench.

Our observations suggest translocation and burial of significant quantities of pre-aged organic carbon in the hadal environment, shedding new light on the nature and dynamics of carbon supply to hadal zone. This has important implications for the identification of gravity flow events triggered by tectonic activity in the Japan Trench sediments, and potentially in other hadal zone sedimentary sequences lying below the CCD.

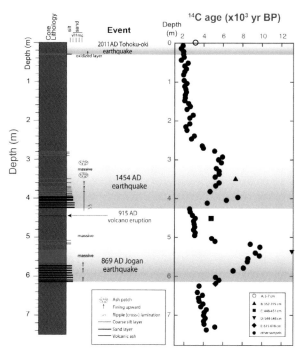

Fig. 1: *Lithology of Core GeoB 16431-1 [1] and a high-resolution ^{14}C age profile of sedimentary organic matter.*

[1]   K. Ikehara et al., Earth Planet. Sci. Lett. 445 (2016) 48

[1] *Geology, ETH Zurich*
[2] *NOSAMS, Woods Hole, USA*
[3] *University of Innsbruck, Austria*
[4] *MARUM, Bremen, Germany*

^{14}C AGE OF GRAIN SIZE FRACTIONS IN THE YELLOW RIVER

Influence of hydrodynamic sorting on organic carbon composition

M. Yu[1,2], T. Eglinton[2] , N. Haghipour[2], D. Montluçon[2], L. Wacker, P. Hou[1] ,M. Zhao[1]

The transport of organic carbon (OC) by rivers to coastal oceans is an important component of the global carbon cycle. The Yellow River (YR), the second largest river in China, transports large amounts of particulate organic carbon (POC) to the Chinese marginal seas, with fossil and pre-aged (ca. 1600 yr) OC comprising the dominant components [1]. However, the influence of hydrodynamic processes on the origin, composition and age of POC exported by the YR remains poorly understood, yet these processes likely play an important role in determining OC fate in the Chinese marginal seas. We address this question through bulk, biomarker and carbon isotopic (δ^{13}C and Δ^{14}C) characterization of organic matter associated with different grain size fractions of total suspended particles (TSP) in the YR. Surface TSP samples were collected in the late spring, in summer and during the Water-Sediment Regulation period (WSR, July) of 2015 at Kenli Station (Fig.1). TSP samples were separated into five grain-size fractions (<8µm, 8-16µm, 16-32µm, 32-63µm and >63µm) with a hydraulic instrument for organic geochemical and isotope analysis.

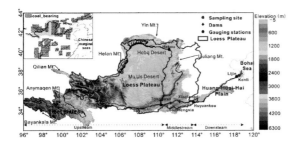

Fig. 1: Elevation map of the Yellow River drainage basin with the sampling station of Kenli (red dot) [1].

Here we report the preliminary ^{14}C results of n-fatty acids among different grain size fractions.

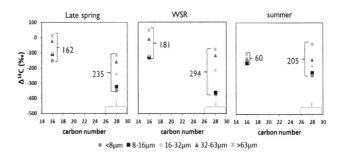

Fig. 2: The Δ^{14}C of n-fatty acids among different grain size fractions of YR TSP before, during and after WSR. Different color and shape of symbols represent different grain size fractions.

Generally, the 16-32µm and 32-63µm fractions contributed most of the TSP mass and the majority of OC resided in 16-32µm fraction. TOC% decreased with increasing grain size and ^{14}C ages exhibited significant variability but did not show any systematic trend among grain size fractions or across sampling times. In contrast, compound-specific ^{14}C analysis of long-chain *n*-fatty acids (C_{26-30} FAs) revealed two clear patterns: First, C_{26-30} FAs age decreased with increasing grain size for all sampling times. Second, the C_{26-30} FAs age difference was the largest among the different size fractions during the WSR period, and smallest after the WSR. These findings have important implications for our understanding of riverine POC transport mechanisms and their influence on the dispersal and burial efficiency of terrestrial OC in coastal oceans.

[1]   S. Tao et al., Earth Planet. Sci. Lett. 414
       (2015) 77

[1] *Ocean University of China, Qingdao*
[2] *Geology, ETH Zurich*

SOIL ORGANIC MATTER VULNERABILITY

The vulnerability of soil organic carbon and its relation to drivers

B. González-Domínguez[12], P.A. Niklaus[2], M.S. Studer[1], F. Hagedorn[3], L. Wacker, N. Haghipour[4], S. Zimmermann[3], L. Walthert[3], C. McIntyre[4,5], S. Abiven[1]

With climate change, the soil organic carbon (SOC) pool is vulnerable to loss as CO_2 [1]. The objective of this project is to investigate the variation of indicators of SOC vulnerability (e.g. SOC mineralisation, turnover time, bulk soil and mineralised carbon ^{14}C signatures) and to evaluate climate (i.e. proxies to soil temperature and moisture), soil (i.e. pH, % clay) and terrain (i.e. slope gradient, orientation) as primary drivers. The approach we developed to select the 54 study sites (Fig. 1) overcame the difficulty of statistically distinguishing between the effects of confounding drivers.

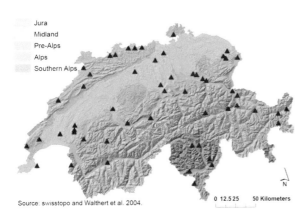

Fig. 1: *Sites were selected from a soil database of over 1050 profiles across Switzerland and managed by the Swiss Federal Institute for Forest, Snow and Landscape Research (WSL). All sites were re-sampled in summer 2014.*

In this project, ^{14}C is a powerful tool to investigate SOC dynamics. To make use of this potential, we incubated soils and quantified SOC mineralisation by trapping the evolved CO_2 in NaOH [2] (Fig. 2). At LIP, a method was developed to efficiently measure the ^{14}C signature of the C trapped in the NaOH. This method consists on the acidification of the traps with H_3PO_4 and the direct measurement of the gas in MICADAS.

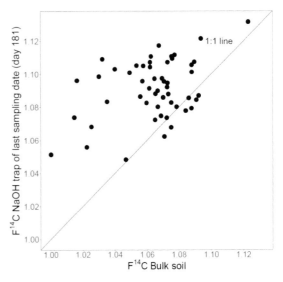

Fig. 2: *Comparison of $F^{14}C$ of bulk soil and mineralised carbon. Sites below the 1:1 line contain more radiocarbon in the bulk soil than in the mineralised fraction.*

Based on the cumulative C mineralised, we found that temperature did not emerge as a predictor of SOC vulnerability. The strongest driver we identified was soil pH. Soils with higher pH also showed higher vulnerabilities. This could be explained by lower pH values favouring stronger bond types [3].

[1] Y. Luo, A. Ahlström, et al., Global Biogeochem. Cycles. 30 (2016) 40.

[2] A. Wollum, J. Gomez, Ecology. 51 (1970) 155.

[3] W.H. Yu, N. Li, et al., Appl. Clay Sci. 80–81 (2013) 443.

[1] *Department of Geography, University of Zurich*
[2] *Department of Evolutionary Biology and Environmental Studies, University of Zurich*
[3] *Forest Soils and Biogeochemistry, WSL*
[4] *Geology, ETH Zurich*
[5] *AMS Laboratory, SUERC*

EXTRANEOUS CARBON IN SMALL-SCALE ^{14}C ANALYSIS

Application of the model of constant contamination

U.M. Hanke[1], L. Wacker, N. Haghipour[2], M.W.I. Schmidt[1], T.I. Eglinton[2], C.P. McIntyre[2,3]

Compound-specific radiocarbon (^{14}C) analyses can provide precise information on the isotopic signature of individual molecules in environmental samples. However, the chemical pre-treatment and purification via preparative chromatography often yields less than 50 µg C including extensive sample transfers as well as exposure to various reagents. This makes such analyses prone to extraneous carbon and can have a large affect on the results. Generally, the impact of the contaminant increases with decreasing sample size. Wacker and Christl [1] suggested the model of constant contamination to determine the mass and ^{14}C signature of the contaminant in a sequence of samples and subsequently enable a correction of measured ^{14}C data via the Pearson chi-squared test.

We assessed the impact of extraneous carbon on small-scale ^{14}C analysis of the combustion markers benzene polycarboxylic acids (BPCAs) with multiple processing standards of known ^{14}C contents. We measured different materials that span over two, three and four sub-procedures in concentration from 10 to 60 µg C (see Fig. 1).

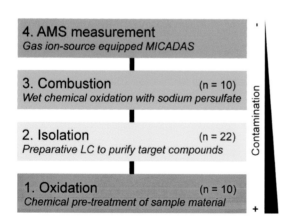

Fig. 1: *Multi-step scheme of extraneous carbon assessment in ^{14}C-BPCA analysis.*

Following the concept of Wacker and Christl (2012) we evaluated ^{14}C depleted and modern

standard materials and fitted modeled curves to the measured data. Results from the model of constant contamination suggest a high statistical confidence for calculations of modeled and measured F^{14}C (Fig. 2).

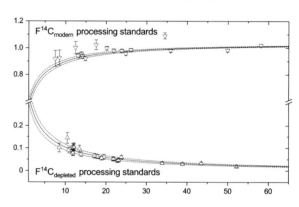

Fig. 2: *Measured ± total uncertainty (symbols) vs. modeled ± 1-σ (lines) F^{14}C of sub-procedures.*

Regardless of the quantity of contamination the statistical fit of processing standards for different sub-procedures remained the same. Nonetheless, the ^{14}C depleted standards show a slightly better fit compared with ^{14}C modern materials indicating a larger variability for fossil contamination.

A thorough quantification of contamination was achieved for molecular ^{14}C data with gas ion-source equipped MICADAS. We strongly suggest to always include sets of processing standards in small-scale molecular ^{14}C analyses.

[1] L. Wacker and M. Christl, ETHZ LIP Ann. Rep. (2012)

[1] *Geography, University of Zurich*
[2] *Geology, ETH Zurich*
[3] *AMS laboratory, SUERC, East Kilbride, UK*

^{14}C CHARACTERISTICS OF BIOMARKERS IN SOILS

Clay content and minerals control long chain FAs and DAs dynamics

J. Jia[1,2], T. Eglinton[2], N. Haghipour[2], L. Wacker, X. Feng[1]

Understanding and predicting the fate of organic carbon (OC) in soils is a key focus to understand the impact of global changes on the global carbon cycle. Previous studies evidenced that soil OC has different decomposability and that chemically labile OC can have old ^{14}C ages [1]. However, few studies have compared the ^{14}C age of various soil OC components on a large scale, which may provide important information on the link between the age and turnover of soil OC components to their sources, molecular structures as well as environmental variables.

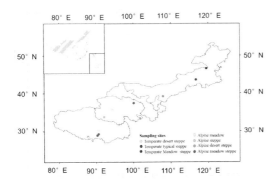

Fig. 1: Sampling sites (total 11) selected along a 3,400 km transect of natural grasslands in China.

Here 11 sites were selected from the temperate and alpine grasslands in the Tibetan-Mongolian of China (Fig. 1). Source-specific compounds were radiocarbon-dated to investigate the age and turnover dynamics of different OC pools and the mechanisms controlling their stability.

Our results show that short-chain fatty acids ($C_{16, 18}$ FAs) sourced from vascular plants as well as microorganisms were younger than plant-derived long-chain FAs and diacids (DAs), indicating that short-chain FAs were easier to be decomposed or newly synthesized. In the temperate grasslands, long-chain DAs were younger than FAs, while the opposite trend was observed in the alpine grasslands (Fig. 2). Preliminary correlation analysis suggests that

the age of short-chain FAs were mainly influenced by clay contents and climate, while reactive minerals, clay or silt particles were important factors in the stabilization of long-chain FAs and DAs (Tab. 1).

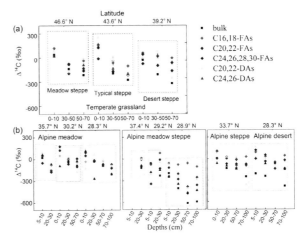

Fig. 2: (a) ^{14}C of bulk, FAs and DAs in temperate grasslands and (b) in alpine grasslands.

	Lat.	Lon.	Al	Clay	Silt	Sand	Alo	Fed	Mnd
Bulk OC	.68	.57	.41	-.18	-.52	.48	-.47	-.64[*]	-.41
$C_{16,18}$ FAs	.68[*]	.62[*]	.61[*]	-.57[*]	-.73[**]	.72[**]	-.26	-0.46	-.59
$C_{20,22}$ FAs	-.01	-.12	.21	-.41	-.78[**]	.75[**]	-.62[*]	-.87[**]	-.84[**]
C_{24-30} FAs	.26	.23	.50	-.60[*]	-.76[**]	.76[**]	-.59	-.79[**]	-.84[**]
$C_{20,22}$ DAs	.55	.49	.43	-.59[*]	-.80[**]	.78[**]	-.68[*]	-.85[**]	-.86[**]
$C_{24,26}$ DAs	.50	.47	.51	-.47	-.84[**]	.81[**]	-.72[*]	-.91[**]	-.79[*]

Tab. 1: *Correlation analysis among ^{14}C of bulk OC and lipids with latitude (Lat.), longitude (Lon.) aridity index (AI), clay content and soil minerals. *: significant at 0.05, **: 0.01 level.*

[1] M. Schrumpf et al., Biogeosciences 10 (2013)

^{1} *Institute of Botany, CAS, China*
^{2} *Geology, ETH Zurich*

TERRESTRIAL CYCLING OF ORGANIC MATTER

Assessing organic matter dynamics in the Godavari River Basin, India

M.O. Usman[1], F. Kirkels[2], H. Zwart[2], S. Basu[3], M. Lupker[1], F. Peterse[2], N. Haghipour[1], L. Wacker, T. Eglinton[1]

Sediment and soil organic carbon (OC) represents a substantial portion of the carbon found in the terrestrial ecosystem of the planet. Despite its relatively small size compared to the oceanic pool, the rate of exchange between the terrestrial pool and the atmosphere is estimated to be higher than that between the ocean and the atmosphere. Rivers serve as a major conduit by which terrestrial-derived organic matter (both in particulate and dissolved form) is cycled within a river basin and between basin and ocean where it is eventually buried. Therefore, a detailed understanding of the composition, provenance and associated timescales of carbon cycling within a basin is crucial in the evaluation of the global carbon cycle (Fig. 1).

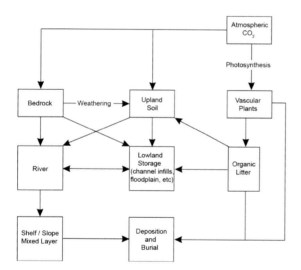

Fig. 1: A conceptual model for the transport of terrestrial OC [1].

Stable and radioactive isotopes of carbon are a novel tool for assessing and characterizing the composition and cycling of organic matter. In this study, we used the bulk geochemical, stable isotope and radiocarbon analysis of sediment and soil samples collected during a dual season sampling campaign in 2015 to explore the source and transport of organic matter. Results show a spatiotemporal heterogeneity in the composition of the OC (Fig. 2).

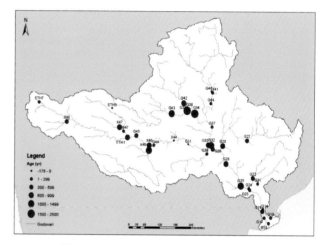

Fig. 2: ^{14}C distribution in the Godavari River Basin.

The upper catchment of the basin shows relatively old radiocarbon ages, the oldest ages are recorded in the sediments from the Prahinta River (a major tributary of the Godavari). We attribute this to open coal-mining activities within this region of the basin. The stable isotope values show a marked contrast between the upper and lower catchment, reflecting changes in the vegetation types from the different source regions as a consequence of aridification of the Indian Peninsula [2].

[1]   Blair et al., Mar. Chem. 92 (2004) 141
[2]   Ponton et al., Geophys. Res. Lett. 39 (2012) 1

[1] Geology, ETH Zurich
[2] Earth Science Department, Utrecht University, Netherlands
[3] IISER, Kolkata, India

REVISITING MILLENNIAL-SCALE CLIMATE VARIABILITY

Asynchronous climate signals in the Shackleton Sites

B. Ausín[1], C. Magill[1,2], N. Haghipour[1], L. Wacker, T. Eglinton[1]

Marine sediment cores from the benchmark Shackleton Sites (Fig. 1) record millennial-timescale climate variability at high resolution. Assessment of such short-term climate variations requires accurate time control of proxy records.

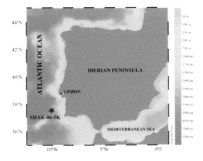

Fig. 1: *Study area and location of the sediment core SHAK-06-5K in the Shackleton Sites.*

In most paleoceanography studies on the last deglaciation and Holocene, a single age-depth model based on ^{14}C dates of foraminiferal shells is applied to all proxy records derived from the same core, assuming that co-occurring sediment components that carry the proxy information (e.g. foraminifera, alkenones, coccoliths, etc.) are coeval.

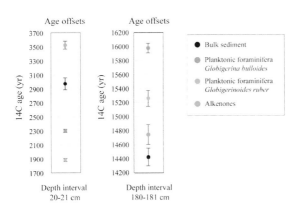

Fig. 2: *Radiocarbon offsets among different carbon-bearing sediment components from the same depth interval.*

Here we report radiocarbon age offsets between bulk sediment, foraminifera, and alkenones of up to several thousand years (Fig. 2), which are tentatively interpreted in terms of hydrodynamic processes and depositional setting. These results evidence that our understanding of rapid climate changes can be highly improved by establishing independent chronostratigraphies for the corresponding proxies (i.e. total organic carbon, δ^{18}O and sea surface temperature). A highly novel aspect of this project is the use of flow cytometry for the isolation and purification of fossil coccoliths (Fig. 3) —a proxy for primary productivity— for subsequent ^{14}C analysis.

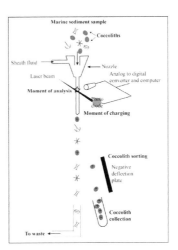

Fig. 3: *Schematic overview of typical flow cytometer sorter.*

This study will yield first independent chronologies for a suite of different proxies that trace surface ocean properties in the Shackleton Sites.

[1] Geology, ETH Zurich
[2] Lyell Centre, Heriot-Watt University, UK

ORGANIC CARBON EXPORT DURING EXTREME EVENTS

Observing the shift of carbon signatures with varying discharge

H. Gies[1], S. Wick[2], M. Lupker[1], C. Freymond[1], F. Peterse[3], N. Haghipour[1], L. Wacker, T. Eglinton[1]

The riverine export of biospheric carbon to the ocean, where it is eventually deposited in marine sediment, is an important long-term carbon sequestration mechanism. High discharge events appear to generate especially high export fluxes of particulate organic carbon (POC). These events mobilize larger proportions of modern organic carbon over petrographic carbon in comparison to background discharges, probably due to topsoil erosion [1]. However, detailed information on the mechanisms of organic carbon mobilization in the stream network is still scarce.

This project aims to show the shift in origins of the exported carbon using biomarkers: the molecular signal of POC collected during high discharges in three alpine headwater streams in the Alpthal (Schwyz, Switzerland, Fig. 1) is compared to possible sources including bedrock, soil and vegetation.

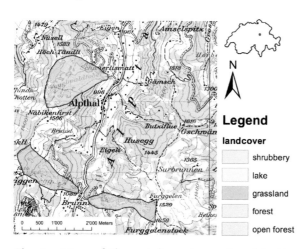

Fig. 1: *Map of the catchments analyzed in the study.*

The analysis indicates that the molecular signature of POC is indeed dependent on discharge (Q) reflecting changing inputs from soil, bedrock and vegetation.

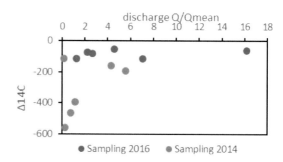

Fig. 2: *Changes in radiocarbon signature in the Lümpenenbach.*

As the respective radiocarbon measurements were conducted in December 2016, the results presented here are only preliminary. First, the radiocarbon signal seems to be dependent on hydrology with a trend towards younger material at higher discharges (Fig. 2).

While the rivers Vogelbach and Lümpenenbach feature POC with a mean $\Delta^{14}C$ of -178.1, the material in the Erlenbach is distinctly older with a mean $\Delta^{14}C$ of -338.1 indicating a higher importance of bedrock-derived carbon in this catchment.

[1] J. Smith et al., Earth Planet. Sci. Lett. 365 (2013) 198

[1] Geology, ETH Zurich
[2] Department of Water Resources and Drinking Water, EAWAG, Switzerland
[3] Department of Earth Sciences, Utrecht University, Netherlands

COSMOGENIC NUCLIDES

Abrasion rates at Goldbergkees, Austria

Dating the stabilisation of relict rockglaciers

Lateglacial history of the high Tatra Mountains

Reconsidering Alpine Lateglacial stratigraphy

Glacier fluctuations at the Mer de Glace

First ^{36}CL ages from a northern Alpine moraine

Lateglacial stages in northern Apennines (Italy)

Dating cirque glacier moraines, Ötztal MTS

Glacier-derived climate for the Younger Dryas

Glaciations in eastern Anatolian Mountains

Glacial stages in the Chagan Uzun Valley, Altai

Isocron-burial dating of Deckenschotter: Siglisdorf

Ages of glacial and fluvial deposits in Patagonia

A ~800 Ka record of terraces and moraines

The Stadlerberg Deckenschotter outcrops: age and evolution

Surface exposure dating in Victoria Land

Post-LGM landscape evolution around Sedrun

Slip rates in western Turkey

1046 AD quake and co-seismic Castelpietra event

Dating the Pragser Wildsee rock avalanche

The Akdag active landslide (SW Turkey)

Estimation of water reservoir lifetimes from ^{10}Be

Tracing the solar 11-year cycle back in time

A new ETH in situ cosmogenic ^{14}C extraction line

Sediment connectivity drives denudation rates

Constraining erosion rates in semi-arid regions

Vegetation modulates denudation rates

The pattern of background erosion rate

Tracing landscape evolution of the Sila Massif

ABRASION RATES AT GOLDBERGKEES, AUSTRIA

Subglacial erosion determined with ^{10}Be and ^{36}Cl

C. Wirsig, S. Ivy-Ochs, J. Reitner[1], M. Christl, C. Vockenhuber, M. Bichler[1], M. Reindl[2], H.-A. Synal

Fig. 1: *Sampling sites at Goldbergkees. Glacier is on upper left (2012).*

We report concentrations of cosmogenic ^{10}Be and ^{36}Cl used to determine erosion depths in the recently deglaciated bedrock at Goldbergkees (Fig. 1) in the Tauern region of the Eastern Alps [1]. The glacier covered the sampling sites during the Little Ice Age (LIA) until ca. 1940. The youngest ages calculated from these concentrations match the exposure time after the known post-LIA exposure of <100 years. The apparent age (assuming no surface cover and no erosion) of most samples, however, is significantly older.

We show that the measured nuclide concentrations represent subglacial erosion depths, rather than exposure times. In particular, erosion depths calculated using ^{10}Be and ^{36}Cl concentrations of individual samples match well, whereas apparent ^{36}Cl ages are consistently older than ^{10}Be ages. This reflects the higher production of ^{36}Cl deeper into rock in comparison to ^{10}Be. The bedrock at the 'youngest' surfaces was deeply eroded (≥297 cm) by the Goldbergkees glacier during the late Holocene. In contrast, bedrock at the margin of the LIA ice extent was eroded ≤35 cm. These values convert to subglacial erosion rates on the order of 0.1 mm/a to 5 mm/a. While modeled erosion rates depend on the duration of glacial cover and erosion intrinsic to the different exposure scenarios used for calculation

(700-3300 years), modeled total erosion depths are insensitive (5-20 % change).

Analysis of erosion depths on the transverse valley profile shows a general trend of greatest erosion part way up the valley side and less erosion under thin ice at the lateral margin. A second profile along the valley axis indicates depth of erosion is greatest where the ice abuts the foot of the investigated bedrock riegel and at its lee side just beyond the crest (Fig. 2).

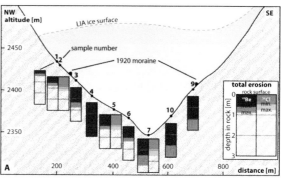

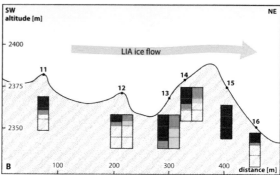

Fig. 2: *Depths of subglacial erosion determined with ^{10}Be and ^{36}Cl.*

[1] C. Wirsig et al., Earth Surf. Proc. Landf. (2017) in press

[1] *Geological Survey of Austria, Vienna, Austria*
[2] *Environmental Geosciences, University of Vienna, Austria*

DATING THE STABILISATION OF RELICT ROCK GLACIERS

Using cosmogenic nuclides to date relict rock glaciers

O. Kronig , J.M. Reitner[1], M. Christl, S. Ivy-Ochs, H.-A. Synal

Active rock glaciers are periglacial landforms that creep downhill. Relict rock glaciers, however, lost their interstitial ice and stabilised. Therefore, the positions of relict rock glaciers give evidence of past discontinuous permafrost distribution [1]. Dating the time of the stabilisation can be used to reconstruct past local permafrost distribution and paleoclimate.

The Tandl rock glacier in the Reisseck mountain group (Carinthia, Austria) is a complex series of relict rock glaciers (Figs. 1&2). The highest lobe is at 2300 m (a.s.l.) and the lowest lobe reaches down to 1220 m a.s.l. According to Reitner [2] this is the lowest relict rock glacier of the Eastern Alps. Due to its extremely low position, it is thought that the rock glacier was active during the Lateglacial.

Fig. 1: *One of the Tandl rock glaciers. The two persons give a reference height of the landform. For the exact location of the picture see Fig. 2.*

A total of 20 samples of the entire succession was taken for ^{10}Be exposure dating. First results show that the lowest rock glaciers indeed stabilised during the Lateglacial. Furthermore, different phases of rock glacier activity were detected as lobes at higher elevation have younger ages.

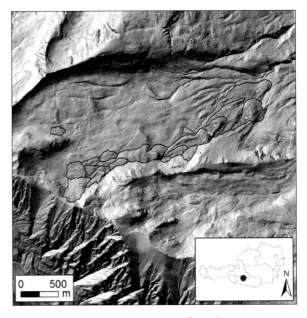

Fig. 2: *Hillshade map (Land Kärnten– data.ktn.gv.at) of the Tandl rock glaciers, highlighted in violet. Blue dots show the sample locations and the orange arrow marks the position and view direction of Fig. 1.*

No farther than 10 km, a second series of rock glaciers is located. In contrast to the Tandl rock glaciers which are facing north-east, these ones are facing south-west. Here, additional 14 samples were collected with the intention to assess the influence of aspect on the permafrost distribution.

[1]   D. Barsch, PPP 3 (1992) 175

[2]   J.M. Reitner, Jahrbuch der Geologischen Bundesanstalt 3-4 (2007) 672

[1] *Geological Survey of Austria, Vienna, Austria*

LATEGLACIAL HISTORY OF THE HIGH TATRA MOUNTAINS

Geomorphological mapping and ^{10}Be dating of glacial landforms

E. Opyrchał[1], S. Ivy-Ochs, J. Zasadni[1], P. Kłapyta[2], M. Christl

The High Tatra Mountains constitute an example of a mountain range with no contemporary glaciation but containing moraines and relict rock glaciers which record the activity of glaciers during the late Pleistocene. Recent investigations were focused on the extent and chronology of the latest glacier advances and have broadened our knowledge in the field of glacier fluctuations. Nevertheless, questions about when the Tatra Mountains have become deglaciated remain without unequivocal answers and need further examination. In this study, Veľká Studená Valley located on the southern slope of the High Tatra Mountains has been chosen as a case study. The upper part of the valley is one of the highest elevated cirques within this mountain range, with the cirque floor at around 1950 m a.s.l., surrounded by steep, several hundred meters high rock walls. The cosmogenic ^{10}Be surface exposure dating method was supported by the Schmidt hammer relative dating method and combined with geomorphological approach. Samples from both boulders and rock surfaces moulded by the ice were collected for dating using cosmogenic nuclides and also tested with the Schmidt-hammer tool to obtain an overall insight into the relative chronology between the landforms.

First results reveal that in the highest parts of the Tatra Mountains, two moraine systems associated with two Lateglacial stages can be found. These glacier advances built most of the landforms which create outstanding landscape of the valley. The older stage represented by a voluminous, almost 10 m high moraines, seems to have been supplied by several rock slope failures (Fig. 1.) Within the younger system, an interesting cross-cutting relations can be observed e.g. where moraine continues under the present level of the lake (Fig. 2.).

Fig. 1: *Northern part of the investigated valley with the voluminous, almost 10 m high moraine, pronounced with the pattern of dwarf pines.*

Fig. 2: *Cross-cutting relations of the moraine and lakes in the upper part of the valley.*

The deglaciation of the Tatra Mountains most likely occurred at the end of the Younger Dryas (around 11.7 ka ago). Results lead to the still open discussion if in the study area the glaciers have ever readvanced during the Holocene.

---

[1] *Geology, AGH University, Kraków, Poland*
[2] *Geography, Jagiellonian University, Kraków, Poland*

RECONSIDERING ALPINE LATEGLACIAL STRATIGRAPHY

Implications of the Lienz area record (Tyrol/Austria)

J.M. Reitner[1] S. Ivy-Ochs, R. Drescher-Schneider[2], I. Hajdas, M. Linner[1]

The sedimentary and morphological evidence for Lateglacial glacier fluctuations in the Lienz area provides a strong case against the currently used five-part stratigraphic subdivision of the Alpine Lateglacial (ca. 19–11.7 ka) comprising the early Lateglacial phase of ice decay, Gschnitz, Senders, Daun and Egesen stadials. The results of a comprehensive geological mapping supported by geochronological data and pollen analysis revealed that the Lateglacial record of the Schobergruppe mountains and the Lienz Dolomites can be subdivided into four unconformity-bounded (allostratigraphic) units which are linked to three climatically and stratigraphically defined phases of glacier activity. Delta deposits and till of local glaciers document the phase of ice decay after the Alpine LGM. Between this period and the Bølling/Allerød (B/A) interstadial only one phase of glacier stabilization, where massive end moraines formed, that can be correlated to the Gschnitz stadial, is evident.

Fig. 2: *Gneiss boulder of sample DEBANT 2 on top of the Egesen latero-frontal moraine in Fig. 1 which was previously regarded as a typical Daun moraine (pre B/A interstadial).*

Multiple end moraines prove the presence of very active glacier tongues during the Younger Dryas-aged Egesen stadial (Fig. 1). The [10]Be exposure dating of an end moraine, previously attributed to the Daun stadial (pre-B/A interstadial) based on ΔELA (equilibrium line altitude) values, provided an age of 12.8 ± 0.6 ka indicating it is of Younger Dryas age (Fig. 2).

This case highlights the pitfalls of the commonly used ΔELA-based stratigraphic Alpine Lateglacial subdivision and the subsequent derivation of palaeoclimatic implications. ΔELAs are still considered as a useful tool for correlation on the local scale, e.g. within one mountain group with a comparable topography and lithology and taking into account the limitations, especially the impact of debris cover.

[1]   J.M. Reitner et al., Quat. Sci. J. 65 (2016) 113

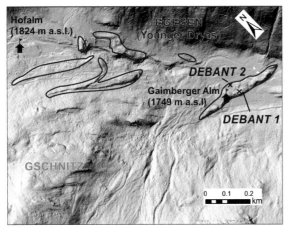

Fig. 1: *Shaded relief image from TIRIS (www.tirol.gv.at) of the Debant Valley with a typical set of "fresh" multiple latero-frontal moraines of the Egesen stadial compared to an isolated moraine of the Gschnitz stadial.*

[1] *Geological Survey of Austria, Vienna, Austria*
[2] *Schillingsdorfer Strasse 27, Kainbach, Austria*

GLACIER FLUCTUATIONS AT THE MER DE GLACE

Combining optically stimulated luminescence and [10]Be exposure dating

B. Lehmann[1], P.G. Valla[2], G.E. King[2], S. Ivy-Ochs, O. Kronig, M. Christl, F. Herman[1]

Providing tight spatial and temporal constraints on late Pleistocene glacier fluctuations remains an important challenge for understanding glacier and glacial erosion responses to climate change. Palaeoglacier reconstructions are often scarce, discrete and spatially limited during the Lateglacial and Holocene times, which makes their use as a paleoclimate proxy sometimes challenging.

Here we focus on the Mer de Glace glacier (Mont-Blanc massif, France) where glacier reconstructions since the middle Holocene [1] reveal important glacier fluctuations and ice thickness variations. However, continuous records of the Mer de Glace fluctuations must be precisely constrained to provide a valuable record of local climate and erosion with time. Therefore, we collected samples of polished granitic bedrock surfaces between the Last Glacial Maximum (LGM) ice surface (~2505 m a.s.l, [2]) and the present-day glacier (1920 m a.s.l) covering ~600 m of elevation for the ice surface fluctuations (Fig. 1).

We used cosmogenic [10]Be exposure dating on quartz to constrain ice surface fluctuations during the Lateglacial and Holocene. Given that the cosmic ray flux produces [10]Be over the first ~3 m below the rock surface, multiple exposure history from complex glacier fluctuations would be difficult to quantify using this chronometer. To improve the temporal resolution for such possible complex exposure histories, we combined cosmogenic [10]Be exposure dating with optically stimulated luminescence (OSL) dating of the rock surface [4]. OSL dating is sensitive to light, based on the progressive bleaching of the OSL signal in a rock sample that depends on its exposure time, mineralogical properties and environmental conditions. Preliminary OSL results from rock slices show increasing exposure age (i.e., deeper bleaching of the OSL signal) with sample elevation. Moreover, our results reveal that the bleaching of the OSL signal is occurring within the first 1-3 cm below the rock surface, potentially offering high resolution dating of the latest exposure following short-lived glacier fluctuations.

[1] M. Le Roy et al., Quat. Sci. Rev. 108 (2015) 1

[2] S. Coutterand and J.F. Buoncristani, Quaternaire 17 (2006) 35

[3] C. Vincent et al., Annals of Glaciology 55 (2014) 15

[4] R. Sohbati et al., Geochron. 38 (2011) 249

Fig. 1: *Mer de Glace (Mont Blanc massif, France) and the two sampled transects, Trélaporte (green stars) and Montenvers (orange dots). Light blue line shows the Last Glacial Maximum reconstruction [2] and the dark blue line, the Little Ice Age glacier reconstruction [3] (Photo : Hagenmuller).*

[1] *Geosciences and Environment, University of Lausanne*
[2] *Geology, University of Bern*

FIRST ^{36}CL AGES FROM A NORTHERN ALPINE MORAINE

Evidence of early Holocene glacier and rock glacier activity

A.P. Moran[1], S. Ivy Ochs, C. Vockenhuber, H. Kerschner[1]

In the Northern Calcareous Alps of Austria a well-preserved moraine system in the Mieminger Range was dated to provide first exposure ages from a region for which hitherto only relative moraine chronologies have been determined (Fig. 1).

Fig. 1: *Location of the investigated site in the Northern Calcareous Alps, Austria.*

The investigated moraine [1] in the north-facing Schwärzkar cirque (Fig. 2), reaches approximately 1 km beyond the extent of the small cirque glacier of the "Little Ice Age" (LIA; modern times). It allows the reconstruction of a palaeoglacier with a surface of about 0.7 km² reaching from 2450 to 1950 m a.s.l. Based thereupon, an equilibrium line altitude depression of -120 m can be calculated referenced to the LIA glacier.

Fig. 2: *Right lateral moraine of the Schwärzkar cirque, Mieminger Range.*

Boulders on the moraine (Fig. 3) were dated with ^{36}Cl to the early Holocene indicating moraine stabilization at ~10.4 ka and climate conditions favorable to glacier formation prior to this time during the Preboreal period.

Fig. 3: *Sampling of a limestone moraine boulder.*

On the proximal side of this moraine, a number of additional boulders were dated in connection with several small relict rock glaciers in the cirque. Their ages are considerably younger and average around 9 ka. These results point to a phase of prolonged instable conditions subsequent to the retreat of the glacier. This suggests the formation of discontinuous permafrost and periglacial activity within the cirque reaching into the Boreal period. All these landforms lie significantly up valley from a series of lateral moraine segments related to the "Egesen" stadial (Younger Dryas cold phase). The ^{36}Cl ages presented here are the first exposure ages gained from moraines in the northern Alps and form a basis for a numerically dated moraine chronology in the northern Alps.

[1] W. v. Senarclens-Grancy, Jahrbuch der Geol. B.A. 88 (1938) 1

[1] *Geography, University of Innsbruck, Austria*

LATEGLACIAL STAGES IN NORTHERN APENNINES (ITALY)

First [10]Be exposure ages constrain a multi-phase Lateglacial history

C. Baroni[1,2], G. Guidobaldi[3], M.C. Salvatore[1], S. Casale[3], M. Christl, S. Ivy-Ochs

The Northern Apennines represent a hinge area between the Mediterranean and continental Europe. Since the end of the Last Glacial Maximum glaciers repeatedly re-advanced during the Lateglacial, experiencing a multi-phase evolution until the end of the Pleistocene, presaging the early Holocene. Glacier positions and relative equilibrium line altitudes (ELA) are supported by geomorphological evidence combined with [10]Be cosmogenic surface exposure dating.

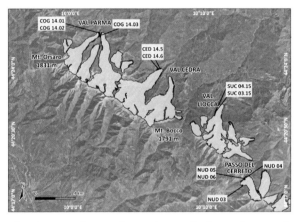

Fig. 1: *Northern Appenines during the early Lateglacial. Moraines in purple. Orthophographs (www.minambiente.it).*

Samples from erratic boulders were collected in selected key sites to outline the glacial history of the mountain range during the Lateglacial. [10]Be exposure ages are obtained for the first time in the Northern Apennines. At least three phases are recorded by the Northern Apenninic palaeoglaciers, constrained by a dataset of [10]Be exposure ages and in high correlation with periodical changes in the North Atlantic circulation and with the consequent climate deterioration periods from the Oldest to the Younger Dryas.

The early Lateglacial stage (Fig. 1) is the response to the Oldest Dryas cold stage and can be related to the Gschnitz Alpine stadial. The second identified phase is tentatively correlated to the Daun Alpine stadial (Fig. 2). Furthermore, [10]Be dates completely support the first direct detection of Younger Dryas moraines in the Northern Apennines (Fig. 2), eventually related to the Egesen Alpine stadial. Moreover, a double response of the Northern Apenninic glaciers was identified related to the Younger Dryas.

Multiphase Lateglacial advances are detected and chronologically constrained for the first time in Northern Apennines. During the Lateglacial cold stages, and in particular during the Younger Dryas, ELAs along the Northern Apennines were at lower elevation with respect to both the Central Apennines and the Alps.

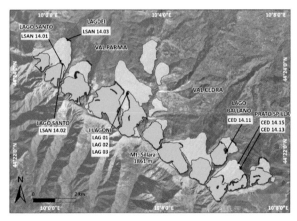

Fig. 2: *Val Parma and Val Cedra during middle and last Lateglacial advances, shown as cyan and green, respectively. Moraines in purple. Orthophotographs (www.minambiente.it).*

[1] *Dipartimento di Scienze della Terra, University of Pisa, Italy*
[2] *C.N.R.-IGG Pisa, Italy*
[3] *Dottorato Regionale in Scienze della Terra, University of Pisa, Italy*

DATING CIRQUE GLACIER MORAINES, ÖTZTAL MTS.

[10]Be dating of glacier extents from the Lateglacial - Holocene transition

A.P. Moran[1], S. Ivy-Ochs, M. Schuh[1], M.Christl, H. Kerschner[1]

Evidence of Alpine palaeoglaciers from the Lateglacial to Holocene transition provides a basis for understanding climate downturns during a time of generally warming conditions. In this context a series of well-preserved and previously undated moraines were investigated in the small Falgin cirque located in the central Alpine Langtaufers Valley (South Tyrol, Italy) and in the neighbouring Hinteres Bergle cirque of the Radurschl Valley (North Tyrol, Austria) [1]. Both localities are situated in the driest area of the Eastern Alps (Fig. 1).

Fig. 1: *Location of the study sites.*

They lie well above prominent moraines associated with the Younger Dryas (YD) cold phase and represent the first moraines below Little Ice Age (LIA) positions [2]. The corresponding equilibrium line altitude of the palaeoglaciers in both cirques was 100-120 m lower than during the LIA. Surface exposure dating with [10]Be of the inner Falgin moraines shows a mean stabilization age of 11.2 ± 0.9 ka, which is similar to the deglaciation age of 10.9 ± 0.8 ka for the Hinteres Bergle cirque (Fig. 2). The ages indicate glacier activity most likely during the earliest Holocene or the YD/Holocene transition. These findings point to a climate with mean summer temperatures about 1.5°C lower than during the 20th century in the Alps. Our results are in agreement with published ages obtained from analogous sites in Switzerland. They are more than 1000 years younger than the much more extensive "Egesen" maximum glacier stadial advance, which is also found at both sites in the research area.

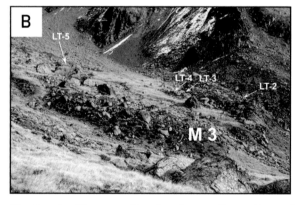

Fig. 2: *A: Hinteres Bergle cirque floor, B: end moraine in Falgin cirque.*

[1]   A.P. Moran et al., Boreas 45 (2016) 398
[2]   H. Kerschner, Alpenvereinsjahrbuch (1982/83) 23 (Z.d.D.u.Ö.A.V. 107)

[1] *Geography, Innsbruck University, Austria*

GLACIER-DERIVED CLIMATE FOR THE YOUNGER DRYAS

A compilation of numerically dated moraine sites in Europe

B. R. Rea[1], R. Pellitero[1], M. Spagnolo[1], J. Bakke[2], S. Ivy-Ochs, P. Hughes[3], R. Braithwaite[3], A. Ribolini[4], S. Lukas[5], H. Renssen[6]

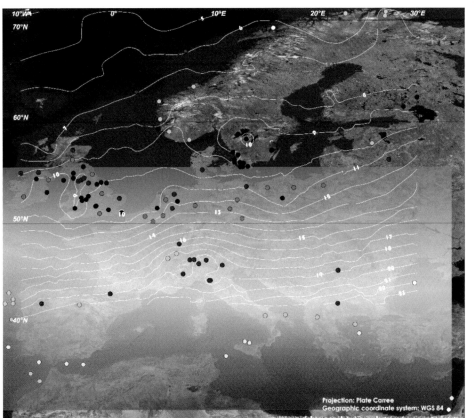

Within this Leverhulme Foundation-funded project we reconstructed and calculated the glacier equilibrium line altitudes (ELA) for 120 Younger Dryas paleoglaciers from Morocco in the South to Svalbard in the North and from Ireland in the West to Turkey in the East. The ages of these landforms were quality controlled; cosmogenic nuclide surface exposure ages were recalculated, where necessary, using currently accepted production rates. Frontal moraines for the paleoglaciers were used as the basis for paleoglacier reconstructions, which were accomplished using the GIS tool we developed [1]. From the resulting equilibrium profiles, Younger Dryas ELAs were calculated. In order to utilize these ELAs for quantitative paleo-precipitation reconstructions a database of 121 sites with independent temperature determinations was compiled. These proxy data, were merged and interpolated to generate maps of average temperature for the warmest and coldest months and annual average. From these maps, the temperature at the paleo ELA was obtained using a lapse rate of 0.65°C/100m.

[1] R. Pellitero et al., Computers and Geosci. 94 (2016) 77

[1] *Geography, University of Aberdeen, U.K.*
[2] *Earth Sciences, University of Bergen, Norway*
[3] *Geography, University of Manchester, U.K.*
[4] *Scienze della Terra, Università di Pisa, Italy*
[5] *Geography, University of London, U.K.*
[6] *Earth Sciences, VU University Amsterdam, NL*

GLACIATIONS IN THE EASTERN ANATOLIAN MOUNTAINS

Surface exposure dating of glacial deposits with ^{36}Cl

S. Yeşilyurt[1], U. Doğan[2], N. Akçar[1], V. Yavuz[3], S. Ivy-Ochs, C. Vockenhuber, F. Schlunegger[1], C. Schlüchter[1]

The chronology of Late Quaternary advances in the northern and western Turkish mountains was reconstructed by surface exposure dating. Glaciation of the Anatolian mountains during the global Last Glacial Maximum (LGM; 22.1 ± 4.3 ka [ka: thousand years], *sensu* [1]) occurred 21.9 ± 1.5 ka ago within the MIS2 [2]. Whereas, glacier advances in eastern Turkey are not dated yet. In this study, we investigated paleoglaciations in the Kavuşşahap Mountains, which are located to the south of Lake Van in eastern Turkey (Fig. 1).

The Kavuşşahap Mountains are one of the most extensively glaciated areas in Turkey. Past glacier activity is evidenced by the more than 20 U-shaped valleys. Paleoglaciers covered ca. 200 km². One of the prominent and well-preserved glacial landscapes of Turkey is the Narlıca Valley system (Fig. 2). The longest glacier in the valley reached a length of 17 km. Lateral and terminal moraines in the valley system indicate several glacier advances. To construct a chronology, 39 erratic boulders were sampled for surface exposure dating with cosmogenic ^{36}Cl.

Fig. 2: *Glacial deposits demonstrate several glacier advances in Narlıca Valley.*

Based on the results, we reconstructed the chronology of four glacier advances in the lower part of the valley system. The local LGM occurred at ca. 47 ka. The second advance was during the early global LGM. The third advance was during the global LGM. The fourth advance of glaciers occurred during the Lateglacial.

[1] J.D. Shakun and A.E. Carlson, Quat. Sci. Rev. 29 (2010) 1801

[2] N. Akçar et al., Geol. Soc. London Spec. Pub. (2017) 433

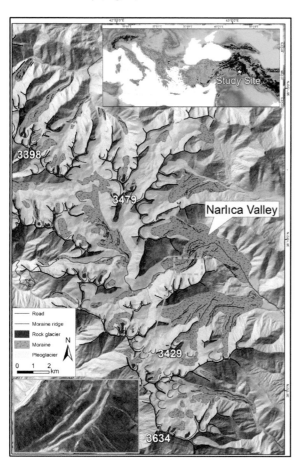

Fig. 1: *Maximum extent of paleoglaciers in the Kavuşşahap Mountains.*

[1] Geology, University of Bern
[2] Geography, Ankara University, Turkey
[3] Geological Engineering, Istanbul Technical University, Turkey

GLACIAL STAGES IN THE CHAGAN UZUN VALLEY

Surface exposure dating of moraines with ^{10}Be in the Altai Mountains

E. García Morabito[1], R. Zech[1], V. Zykina[2], M. Christl

The mountains in southern Siberia (Altai, Sayan, Transbaikalia) were glaciated repeatedly during the Quaternary, and the glacial sediments and landforms there are valuable archives for paleoenvironmental and climate reconstructions. However, studies focusing on glacial chronologies in these mountain ranges are still very scarce, and the extent and timing of glacier advances remain poorly constrained [1], [2].

We have used cosmogenic ^{10}Be surface exposure dating to establish a glacial chronology for the Chagan Uzun Valley in the southern Chuja Range in the Russian Altai. We targeted six moraine complexes and sampled gneiss boulders and quartz cobbles (Fig. 1).

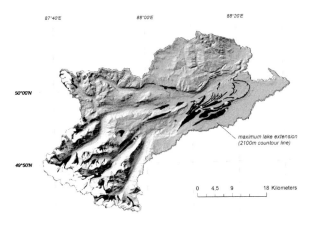

Fig. 1: *Dem (SRTM 30 m, artificially illuminated) of the Chagan Uzun catchment showing principal ice limits, moraines, paleolake levels and sample sites.*

Exposure ages vary widely on the individual moraines, making interpretations in terms of depositions ages and timing of glacier advances difficult. Assuming negligible nuclide inheritance and interpreting the oldest sample from a moraine as best available estimate, massive glaciation occurred as early as ~84 ka. Subsequent, successively less extensive glacier advances probably occurred at >40 ka (MIS4?), and 21-24 ka.

Our data highlight the difficulty to apply surface exposure dating of glacial landforms in cold, continental regions, most likely due to long-lasting landform surface instability. The wide scatter in ages also reflects the highly dynamic context of the region, where glacier advances coexisted with ice-dammed lakes and cataclysmic flood events.

Our results nevertheless add additional evidence for an early local Last Glacial Maximum (LGM) in Siberia, significantly predating the global LGM. Deglaciation and the last massive flood probably occurred at ~18.5 ka, as indicated by our exposure ages in the Chibit area and previously published ages [3].

[1]   F. Lehmkuhl et al., Dev. Quat. Sci. 15 (2011) 967
[2]   A. Reuther et al., Geology 34 (2006) 913
[3]   A. Agatova et al., STRATI 2013 (2014) 903

[1] *Geography, University of Bern*
[2] *Geology of the UIGGM, Russia*

ISOCHRON-BURIAL DATING OF DECKENSCHOTTER

Deposition age of the Higher Deckenschotter deposit at Siglistorf

N. Akçar[1], S. Ivy-Ochs, V. Alfimov, F. Schlunegger[1], A. Claude[1], R. Reber[1], M. Christl, C. Vockenhuber,
A. Dehnert[2], M. Rahn[2], C. Schlüchter[1]

We used the isochron-burial technique to date the oldest glacially derived Quaternary units in the northern Alpine foreland (the Deckenschotter), whose chronostratigraphy remains inadequately reconstructed. This requires analysis of both cosmogenic ^{10}Be and ^{26}Al in samples collected from the same horizon in a gravel deposit. In addition to verifying low cosmogenic nuclide concentrations in the glaciofluvial clasts derived from erosion in the source area, we analyzed the relationship between the ^{26}Al/^{10}Be ratio, depth below surface, and rock density in landscapes governed by deep erosion such as glacial landscapes. With this analysis, we demonstrate that the ^{26}Al/^{10}Be ratio can be as high as 8.4 when clasts originate from depths of 5-10 m. Therefore, we consider 6.75 as the lower limit and 8.4 as the upper limit for the slope of the initial isochron line at the time of burial. We processed a total of 22 quartz-bearing clasts from a Higher Deckenschotter (HDS) natural outcrop in Siglistorf (Canton Aargau, Fig. 1) for isochron-burial dating (Fig. 2).

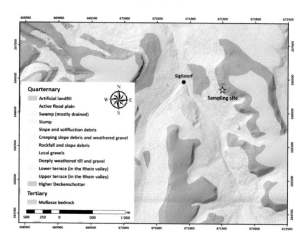

Fig. 2: Close-up view of the Deckenschotter gravels at the Siglistorf site.

After identifying outliers, the remaining six samples were then used to model the upper limit for the isochron-burial age using an initial isochron line with a slope of 8.4 at the time of burial. We calculated an isochron-burial age of 1.7 ± 0.2 Ma. We tested the plausibility of an initial isochron slope of 8.4 by modeling the variation of ^{10}Be, ^{26}Al, and the ^{26}Al/^{10}Be ratio with various exposure times, erosion rates, and local production rates. Taking into account the uncertainty of this age and in consideration with dates determined at two other sites (Irchel and Stadlerberg [2]), we concluded that the HDS at Siglistorf was deposited between 2.5 and 1.5 Ma.

[1] N. Akçar et al., submitted to ESPL
[2] A. Claude et al., Swiss J. Geosci. 107 (2017) 337

[1] Geology, University of Bern
[2] Swiss Federal Nuclear Safety Inspectorate ENSI

AGES OF GLACIAL AND FLUVIAL DEPOSITS IN PATAGONIA

Surface exposure dating with ^{10}Be at the Chile Triple Junction

J. Tobal[1], E. García Morabito[2], C. Terrizzano[2], M. Christl, J. Struck[2], L. Schweri[2], R. Zech[2]

Surrounding the Buenos Aires Lake in Patagonia, at ~46º30′S, there are unusually well-preserved moraine deposits, along with outwash and fluvial terraces that may be climatic counterparts. Both sequences potentially record climatic shifts over the last million years [1].

In the last decade, surface exposure dating of boulders of the youngest moraines (i.e., the Fenix and Moreno moraine systems) has been carried out [2,3]. Nevertheless, ages of the oldest moraines (i.e., the Deseado and Telken moraine systems, Fig. 1) have been constrained so far only by ^{40}Ar-^{39}Ar dating in intercalated lava flows [4]. Surface exposure ages on terraces are still missing.

Fig. 1: *Boulders of the Telken moraine system, east of the Buenos Aires Lake.*

To chronologically constrain the glacial and the fluvial sequence and to confirm or rule out their genetic link, we applied exposure dating with ^{10}Be on glacial boulders at the northern and eastern side of the lake and on outwash and fluvial terraces located further east along the Río Deseado (Fig. 2).

Fig. 2: *Views of the Río Deseado terraces and examples of samples collected for surface exposure dating.*

Our first results indicate an age of ca. 1 Ma for the oldest outwash terraces, which agrees with previous ages on the Telken moraine system. This indicates that outwash and fluvial terraces are possible paleoclimatic indicators in Patagonia.

[1]   A.S. Hein et al., Earth. Planet. Sci. Lett. 286 (2009) 184

[2]   M.R. Kaplan et al., Geol. Soc. Am. Bull. 116 (3/4) (2004) 308

[3]   D.C. Douglass et al., Quat. Geochron. 1 (2006) 43

[4]   B.S. Singer et al., Geol. Soc. Am. Bull. 116 (3/4) (2004) 434

[1] *Institute of Andean Studies, University of Buenos Aires, Argentina*
[2] *Geography, University of Bern*

A ~800 KA RECORD OF TERRACES AND MORAINES

Surface exposure dating with ^{10}Be in the Southern Central Andes

C. Terrizzano[1], E. García Morabito[1], R. Zech[1], M. Christl, J. Likermann[2], J. Tobal[2], M. Yamin[3]

The role of tectonics versus climate in controlling the evolution of alluvial fans is discussed controversially. The southern Central Andes and their forelands provide a perfect setting to study climate versus tectonic control of alluvial fans. On the one hand, the region is tectonically active and alluvial fan surfaces are offset by faults. The higher summits, on the other hand, are glacierized today, and glacial deposits document past periods of lower temperatures and increased precipitation.

We applied ^{10}Be surface exposure dating on five fan terraces and four moraines of the Ansilta range (31.6°S-69.8°W, Fig.1) using boulders and amalgamated pebbles to explore their chronological relationship.

The moraines document glacier advances during cold periods at the marine isotope stages (MIS) 2, 8 and 12. The terraces T1 and T4 seem to be geomorphologic counterparts during MIS 2 and 12.

The other terraces are not represented in the glacial record but the persistence of surfaces of the same age [1] [2] [3] [4] [5] across wide areas in the piedmont suggest aggradation being controlled by a regional driving force such as climate. We then argue that T2, T3 and T5 document aggradation during the cold periods MIS 6, 8 and 18 in response to glacier advances, although the respective moraines are not preserved.

Our results highlight that arid climate in the Southern Central Andes favors the preservation of glacial and alluvial deposits allowing landscape and climate reconstructions back to ~800 ka. Alluvial deposits correlate with moraines or have a regional extension and fall into cold glacial times, so that climate seems to be the main forcing of alluvial fan formation at our study site.

[1] L. Siame et al., Geology 25 (1997) 975
[2] L. Siame et al., Geophys. J. Int. 150 (2002) 241
[3] L. Siame et al., Tectonics 34 (2015) 1129
[4] K. Hedrick et al., Quat. Sci. Rev. 80 (2013) 143
[5] S. Schmidt et al., Tectonics 30 (2011)

[1] Geography, University of Bern
[2] Institute of Andean Studies, University of Buenos Aires, Argentina
[3] Segemar, Buenos Aires, Argentina

Fig. 1: *Moraines and alluvial fans in the Ansilta Range, southern Central Andes.*

THE STADLERBERG DECKENSCHOTTER OUTCROPS

Age and evolution of early Pleistocene glaciofluvial gravels

A. Claude[1], N. Akçar[1], S. Ivy-Ochs, F. Schlunegger[1], P.W. Kubik, A. Dehnert[2], J. Kuhlemann[2], M. Rahn[2], C. Schlüchter[1]

Deckenschotter deposits are considered to be the oldest Quaternary units in the northern Swiss Alpine Foreland. In order to contribute to the understanding of Quaternary landscape evolution in the foreland we applied a multidisciplinary approach at the site Stadlerberg. We dated the Deckenschotter deposits using cosmogenic ^{10}Be depth profile dating and performed sedimentological analyses to identify the provenance of the sediments and to interpret the transport mechanisms and depositional environments.

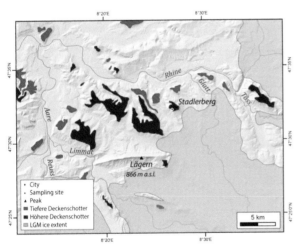

Fig. 1: *Study area Stadlerberg and surrounding Deckenschotter sites [1].*

The site Stadlerberg is an isolated plateau that stands approximately 150 m above the surrounding topography beyond the extent of the Linth Glacier during the Last Glacial Maximum (Fig. 1). We focused our analyses on an abandoned gravel pit on the SSW side of Stadlerberg (Fig. 2).

The obtained results show that these sediments accumulated at minimum at 1.9±0.2 Ma. The source area of the gravels comprised the Aar and Gotthard Massifs, Penninic thrust nappes and Helvetic nappes in the northern Central Alps

as well as the Hörnli and minor parts of the Napf talus fans in the Molasse Basin in the foreland [1]. The sediments were brought to the study site mainly as bedload in a glaciofluvial braided river system. In addition, the results show that the gravels at Stadlerberg were deposited within a distance between a few hundred meters to a few tens of kilometers from the feeding up-valley paleoglacier [1].

Furthermore, we estimated a catchment-wide paleo-denudation rate in the order of 0.3-0.4 mm/a for the study site. It can be concluded that the landscape at ca. 2 Ma ago probably featured a low relief topography with smoother hillslopes than at present.

Fig. 2: *Photograph showing the abandoned gravel pit at Stadlerberg with the locations sampled for ^{10}Be depth profile dating.*

[1] A. Claude et al., Geomorphology 276 (2017) 71

[1] *Geology, University of Bern*
[2] *Swiss Federal Nuclear Safety Inspectorate ENSI*

SURFACE EXPOSURE DATING IN VICTORIA LAND

New evidence of the Late Cenozoic glacial history of Antarctic Ice Sheet

C. Baroni [1,2] S. Casale[3], M.C. Salvatore[1,2], M. Christl, S. Ivy-Ochs

A new set of [10]Be dates was obtained from samples collected during the Antarctic expedition conducted in the Victoria Land in the framework of the Italian National Antarctic Program (PNRA).

The dating of glacial drifts and relict surfaces sampled in key sites of Victoria Land (East Antarctica, Fig. 1) selected on the basis of geomorphological and glacial geological surveys [1] provides new insight into the glacial history and evolution of the Antarctic Ice Sheet. In particular, we obtained new data related to the Last Glacial Maximum (LGM) in southern Victoria Land. New [10]Be dates indicate the permanence on Ross Island of the marine-based Ross Ice Sheet (RIS) at least from 19 to 15 ka. Furthermore, new dates of erratic boulders at low elevations represent minimum ages for RIS persistence on Ross Island until at least 12 ka. Similar dates bracketing the LGM have been obtained in northern Victoria Land. The deglaciation at Terra Nova Bay was still ongoing at ca. 9-8 ka.

A multi-phase evolution of the Younger Drift is detected defining two or even three glacial phases attributable to the late Pleistocene. We identified a phase (> ca. 40 ka) pre-dating the LGM. An older phase can be tentatively constrained to ca. 90 ka. If confirmed, the latter phase can be correlated to the oldest cold period of the early late Pleistocene. Multiple glacial phases characterized the glacial dynamics along the entire Victoria Land. The obtained ages span over the 420-120 ka time interval and constrain different glacial phases that occurred during the Middle Pleistocene.

The multi-nuclide approach ([10]Be, [26]Al) applied on samples collected in the upper basin of the Priestley Glacier support the hypothesis that the Priestley Névé and the East Antarctic Ice Sheet did not override the area after ca 350 ka.

Finally, new dates of relict surfaces in the upper Rennick basin testify a long-lasting exposure to cosmogenic nuclides. The new data, correlated with available ages from other sites [2,3] demonstrate conditions of cold and hyper-arid desert persisting at least since 5-7 Ma.

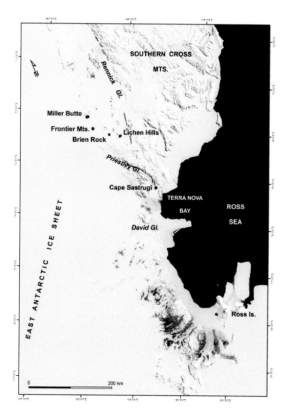

Fig. 1: *Sample location map.*

[1]   C. Baroni et al., Ann. Glaciol. 39 (2004) 256
[2]   P. Oberholzer et al., Antarc. Sci. 15 (2003) 493
[3]   L. Di Nicola et al., Quat. Res. 71 (2009) 83

[1] *Dipartimento di Scienze della Terra, University of Pisa, Italy*
[2] *C.N.R.-IGG, Pisa, Italy*
[3] *Dottorato Regionale in Scienze della Terra, University of Pisa, Italy*

POST-LGM LANDSCAPE EVOLUTION AROUND SEDRUN

Debris-flow fan deposition in the upper Surselva

C. Dielemann, S. Ivy-Ochs, K. Hippe, F. Kober[1], M. Christl

We used geomorphologic field mapping, ArcGIS-based landform interpretation and [10]Be surface exposure dating to reconstruct the Lateglacial to Holocene evolution of the upper Surselva, Graubünden (Switzerland). Four huge debris-flow fans – Pulnanera, Rueras, Drun, Bugnei (from west to east) – dominate the landscape (Fig. 1). Several conflicting theories for the timing of onset and cessation of fan aggradation have been proposed.

Fig. 1: *Alluvial fans in the Sedrun area; P=Pulanera, R=Rueras, D=Drun, B=Bugnei. Note the abruptly cut-off fan toes. Sediment source areas for the Drun and Bugnei fans are visible in the background.*

Based on the depth of bedrock beneath the valley fill gleaned from NEAT tunnel core data, we estimated the total volume of the fans at 35, 54, 106 and 26 million m^3, respectively. However, no dating is available for the cores.

Neither material in the canyons that cut into the fan deposits nor boulders on the fan surfaces proved suitable for dating. Therefore, we chose to constrain maximum ages for the fan surfaces by cosmogenic [10]Be exposure dating of boulders located on moraine ridges that are partially buried by the debris-flow deposits of the fans. Based on our mapped extent of the paleoglaciers in the Giuv, Mila and Strem Valleys

(Fig. 2), we determined their respective equilibrium line altitudes (ELAs) using different ArcGIS tools [1, 2] and both the Accumulation Area Ratio (AAR) and Area Altitude Balance Ratio (AABR) methods. Combining these results with the obtained age data points to moraine construction during the Egesen stadial (~12.9-11.6 ka). The alluvial fans around Sedrun are clearly younger; they formed during the Holocene. The underlying driving factor for abrupt abandonment and subsequent toe cut-off at all four fans is the focus of our study.

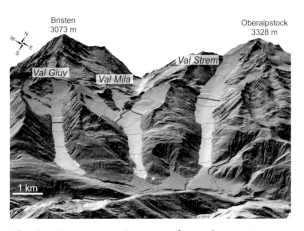

Fig. 2: *Reconstruction of the Egesen paleoglaciers at Sedrun. Pink and yellow lines are ELAs calculated according to the AAR and AABR methods, respectively; yellow dots give the sampling locations for [10]Be exposure dating.*

[1] R. Pellitero et al., Comput. Geosci. 94 (2016) 77
[2] R. Pellitero et al., Comput. Geosci. 82 (2015) 55

[1] *NAGRA, Wettingen*

SLIP RATES IN WESTERN TURKEY

Fault scarp dating with cosmogenic ^{36}Cl

N. Mozafari Amiri [1], Ö. Sümer[2], D. Tikhomirov[1], Ç. Özkaymak[3], B. Uzel[2], S. Ivy-Ochs, C. Vockenhuber, H. Sözbilir[2], N. Akçar[1]

The approximately 140 km long Büyük Menderes graben, formed in response to a roughly N-S extensional regime since the early Miocene, is one of the main tectonic structures in the tectonically active western Turkey. The 37 km long NE-SW trending normal Priene-Sazlı fault in the westernmost part of the graben system is composed of calcareous rocks that rise up to 200 m higher than the Neogene sediments of the Söke-Milet basin (Fig. 1).

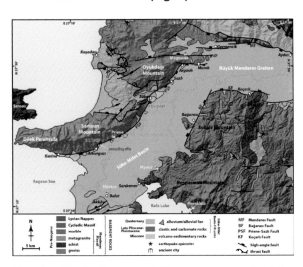

Fig. 1: *Geologic map of the western part of the Büyük Menderes graben [1].*

Two destructive earthquakes were recorded along the Priene-Sazlı fault, one in 68 AD the other in 1955. However, seismic activity prior to this time span is unknown. In order to evaluate the destructive ruptures of the major faults and forecast probable future earthquakes, a complete record of seismic data over a large time scale is required. However, the oldest historical earthquake in the Eastern Mediterranean and Middle East dates back to 464 BC and instrumental earthquake data are only available for the last century. In this study, we investigated the lowest part of the Priene-Sazlı fault scarp close to the ancient city of

Priene (Fig. 2). Collecting 117 rock samples, we applied fault scarp dating with cosmogenic ^{36}Cl concentrations using the FSDT Matlab® code to reconstruct the timing of paleoearthquakes. We modelled five ruptures along the fault with a recurrence interval of about 2000 years. The older ruptures occurred at ca. 4, 6 and 8 kyr ago with slip distances of maximum 3 m. A slip rate of roughly 1 mm/yr is calculated for the active periods. Two of the youngest reconstructed ruptures correlate with the 68 AD and 1955 earthquakes. Taking into account the length of the fault, the capability of producing earthquakes with a magnitude of about 6.8 during single displacement events is probable.

Fig. 2: *Sampled surface of the Priene-Sazlı fault scarp.*

[1]   Ö. Sümer et al., Geodynamics 65 (2013) 148

[1] *Geology, University of Bern*
[2] *Geological Engineerign, Dokuz Eylül University, Turkey*
[3] *Geological Engineering, Afyon Kocatepe University, Turkey*

1046 AD QUAKE AND CO-SEISMIC CASTELPIETRA EVENT

Paleoseismicity underscored with ^{36}Cl exposure landslide dating

S. Ivy-Ochs, S. Martin[1], P. Campedel[2], K. Hippe, C. Vockenhuber, G. Carugati[3], M. Rigo[1], D. Pasqual[1], A. Viganò[2]

Castelpietra, in the Adige Valley (northern Italy), encompasses a main blocky deposit, with an area of 1.2 km^2, which is buried on the upper side by more recent rockfall debris (Fig. 1). ^{36}Cl exposure dates from two boulders in the main deposit indicate an age of 1060 ± 270 AD (950 ± 270 yr ago). The close coincidence in time of the Castelpietra event with several events that lie within a maximum distance of 20 km (Fig. 2), including Kas at Marocche di Dro, Prà da Lago (PdL) and Varini (at Lavini di Marco, LdM) landslides, strongly suggests a seismic trigger.

Based on historical seismicity compilations, we have identified the "Middle Adige Earthquake" at 1046 AD as the most likely candidate. Its epicenter lies right in the middle of the spatial distribution of the four contemporaneous major landslides.

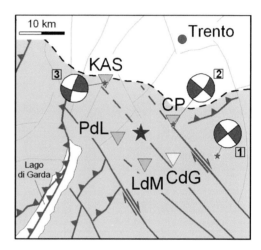

Fig. 2: *Seismotectonic sketch, orange is high-seismicity area, major faults in gray, active faults in magenta, yellow triangle is Cogola di Giazzera (CdG) cave. 1046 AD earthquake epicenter from [2] as blue star. Three recent earthquake epicenters from [3] shown as magenta stars with focal mechanisms.*

[1]   S. Ivy-Ochs et al., World Landslide Forum proceedings (2017) in press
[2]   E. Guidoboni and A. Comastri, Istituto Nazionale di Geofisica e Vulcanologia, Storia Geofisica Ambiente (2005)
[3]   A. Viganò et al., Tectonophys. 661 (2015) 81

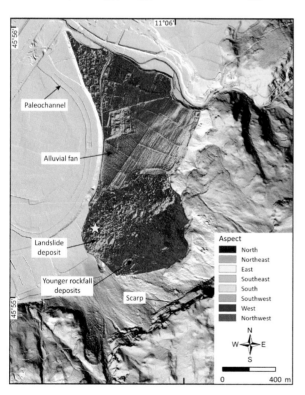

Fig. 1: *Hillshade map (sun azimuth 225°) of the Castelpietra site; landslide and fan deposits are shown with aspect map. Dated boulders at the yellow star. Bedrock shown is Dolomia Principale (pink) and the overlying Calcari Grigi Group (blue).*

[1] *Geoscience, University of Padua, Italy*
[2] *Servizio Geologico della Provincia Autonoma di Trento, Italy*
[3] *University of Insubria, Como, Italy*

DATING THE PRAGSER WILDSEE ROCK AVALANCHE

Surface exposure dating of boulders with ^{36}Cl

M. Ostermann[1], S. Ivy-Ochs, F. Ruegenberg[1], C. Vockenhuber

The Pragser Wildsee/Lago di Braies is a backwater lake of a massive Holocene rock avalanche that still dams the Prags Valley in the northern Dolomites (South Tyrol, N-Italy; Figs. 1&2). The catastrophic rock slope failure involved a rock volume of 30-40 Mm³ with a runout of 8.5 km over a total vertical distance of 1150 m (Fahrböschung 8°). The rock debris now covers an area of about 3.5 km² and dams the 0.4 km² lake. Radiocarbon dates from wood samples on top of the rock avalanche deposits in the lake indicate a minimum age for the slope failure of 7460 ± 100 cal years BP [1].

Fig. 1: *The rock avalanche at Pragser Wildsee indicated on a Lidar-derived hillshade image.*

The Pragser Wildsee rock avalanche deposit and the formation of Pragser Wildsee was previously thought to be of Lateglacial age, however, the present study shows that this is not the case. We applied cosmogenic ^{36}Cl surface exposure dating of four boulders within the debris accumulations and obtained an early Holocene age for the event.

The central and most distal parts of the rock avalanche deposits are characterized by hills and ridges that can be assigned to a toma landscape (Fig. 3), a typical feature for long runout rock avalanches.

Fig. 2: *The Pragser Wildsee is dammed by a massive early Holocene rock avalanche.*

Fig. 3: *The middle part of the Prags Valley is characterised by toma hills associated with the rock avalanche.*

[1] R. Irmler et al., Geomorph. 77 (2006) 69

[1] *Geology, University of Innsbruck, Austria*

THE AKDAG ACTIVE LANDSLIDE (SW TURKEY)

Nature and timing of a complex landslide, using cosmogenic ^{36}Cl

C. Bayrakdar[1], N. Akçar[2] , T. Gorum[1], S. Ivy-Ochs, C. Vockenhuber

Landslides triggered by bedrock failures are one of the main geohazards in high mountain areas. They constitute some of the largest landslide deposits on the Earth. This study focuses on the geomorphological evolution of the Akdag rockslide, which is located on the southern slope of Mount Akdag, SW Turkey (Fig. 1).

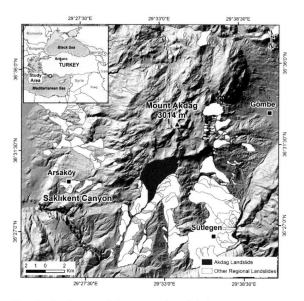

Fig. 1: *Active and inactive landslide areas at the foot of the Akdag Massif, SW Turkey.*

The Akdag Massif is characterized by autochthonous carbonates and shales, which are overridden by allochthonous Jurassic-Cretaceous carbonates. Akdag was glaciated at least three times during the late Pleistocene. There are large karstic depressions developed in these glaciated areas. These karstic depressions are underlain by impermeable flysch deposits. Karstic springs are located along the contact between the flysch and the overlying carbonates.

Our field mapping shows that the Akdag rockslide is a very large and active slope failure developed along the contact between the carbonates and the flysch. Its deposits cover an area of 15 km^2 and have a volume of about 7 km^3. It is one of the largest (>10^6 m^3) landslides in Turkey. Settlements and their infrastructure have been severely damaged due to this activity.

Fig. 2: *Source area of the Mount Akdag landslide.*

In this study, we employed detailed mapping in the field, spatial and morphometric analysis using GIS and remote sensing technologies, and surface exposure dating with cosmogenic ^{36}Cl, in order to reconstruct the chronology of the landslide. For the analysis of cosmogenic ^{36}Cl, we collected 18 samples from calcareous boulders within the landslide deposit. Cosmogenic ^{36}Cl exposure ages indicate that the first collapse occurred at around 9 ka. Based on the field evidence, we anticipate that the increased water discharge in the springs along the carbonate-flysch contact could have played a key role in the failure in the early Holocene.

---

[1] *Geography, Istanbul University, Turkey*
[2] Geology, University of Bern

ESTIMATION OF WATER RESERVOIR LIFETIMES FROM [10]BE

Another application of cosmogenic nuclides

C. Heineke[1], R. Hetzel[1], C. Akal[2], M. Christl

In general, drainage basin erosion and the sediment influx to water reservoirs is determined by river gauging methods, but deduced erosion rates typically have short integration times of a few decades only and are likely influenced by anthropogenic effects [1, 2]. In contrast, catchment-wide [10]Be-based erosion rates can provide robust estimates of drainage basin erosion on millennial timescales that incorporate infrequent erosional events (e.g. debris flows) and are unaffected by short-term or anthropogenic disturbances [1].

To quantify long-term drainage basin sediment yields and provide estimates of operational water reservoir lifetimes, we determined catchment-wide [10]Be erosion rates for the drainage basins of five reservoirs in the central Menderes Massif, Western Turkey (Fig. 1; Tab. 1).

Fig. 1: *Water reservoir in the Menderes Massif.*

Catchment-wide [10]Be erosion rates for these basins vary from 66 to 303 mm/kyr (or 1.7 to 7.6×10^2 t/km^2/yr) and corresponding sediment yields are $10 - 272 \times 10^3$ t/yr (not listed in Tab. 1). By assuming a density of 1.4 ± 0.2 g/cm^3 for the sediment deposited in the reservoirs, we calculated reservoir lifetimes ranging from ~23 to ~680 years until complete filling (Tab. 1). The storage capacities of four

reservoirs will be reduced to 50 % within a few decades, thus severely compromising their long-term operation. We suggest that erosion rates derived from cosmogenic [10]Be are better suited to measure basin erosion and predict reservoir lifetimes than conventional river gauging methods, particularly in regions where water and sediment supply to reservoirs is infrequent and data on sediment yield are not available.

Reservoir Lat.; Long.	[10]Be concen- tration [10^4 at/g]	Erosion rates [10^2 t/km^2/yr]	Volume [10^6 m^3]	Reservoir lifetime [yr]
Çatak 38.29°; 28.16°	2.76 ± 0.30	6.73 ± 0.90	0.68	**46 ± 9**
Başçayir 37.96°; 28.08°	2.05 ± 0.14	7.58 ± 0.77	0.73	**23 ± 4**
Sultanhisar 37.92°; 28.17°	7.01 ± 0.31	2.25 ± 0.20	1.75	**99 ± 17**
Bademli 38.08°; 28.06°	10.25 ± 0.51	1.66 ± 0.15	4.98	**680 ± 120**
Yenişehir 38.07°; 27.92°	2.99 ± 0.18	4.92 ± 0.47	0.72	**98 ± 17**

Tab. 1: *Estimated reservoir lifetimes of five water reservoirs in Western Turkey. Error limits are 1σ.*

[1] J. Kirchner et al., Geology 29 (2001) 591
[2] H. Meyer et al., Int. J. Earth Sci. (Geol Rundsch) 99 (2010) 395

[1] *Geology and Paleontology, University of Münster, Germany*
[2] *Geological Engineering, University of Izmir, Turkey*

TRACING THE SOLAR 11-YEAR CYCLE BACK IN TIME

^{10}Be-based reconstruction of the 11-year solar cycle during Holocene

E. Nilsson[1], M. Christl, J. Beer[2], F. Adolphi[1], S. Bollhalder Lück[2], R. Muscheler[1]

The processes underlying the solar influence on the Earth's climate are a matter of current debate. The study of past solar activity can provide crucial information for a better understanding of the sun-climate link. The most prominent feature of solar variability is the 11-year solar cycle. The length of this cycle varies and has been suggested to be connected to the level of solar activity. The proposed relationship is that higher solar activity leads to shorter cycles while lower solar activity is characterized by longer cycle lengths. The 11-year solar cycle can be seen in the longest observational record of solar activity - the sunspot record [1] (Fig. 1). Using, for example, the wavelet transform it is possible to trace the length of the 11-year solar cycle in the sunspot data through time (Fig. 2).

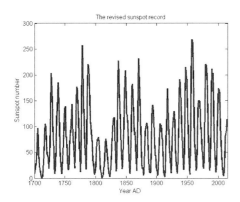

Fig. 1: *The recently calibrated international sunspot number from 1700 until now (source: WDC-SILSO, Royal Observatory of Belgium, Brussels, [1]).*

The Greenland Ice core Project, GRIP, recovered a more than 3 km long ice core in the early 1990s. For the Holocene period, a large part of the ice core has been analyzed for ^{10}Be [2, 3] with an average resolution of approximately 4.5 years. However, between 4100 and 7800 BP the ice core was prepared for high-resolution ^{10}Be measurements of which only every second

sample has been measured until now. In collaboration with the group in Orsay, France, (PI: Grant Raisbeck) this project aims at measuring the remaining samples to increase the average resolution of the GRIP ^{10}Be record to approximately 2.6 years for this period. This will enable us to study the 11-year solar cycle back through time. Hence, we will be able to investigate the proposed connection between the solar cycle length and the solar activity level.

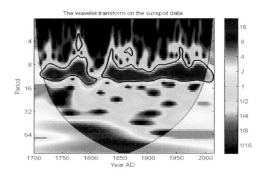

Fig. 2: *The 11-year solar cycle length inferred from the sunspot data with a wavelet transform spectrum.*

Half of the remaining samples between 4100 and 7800 BP from the GRIP ice core are being prepared at the new ^{10}Be preparation laboratory at the Geology Department, Lund University and will be measured ETH Zurich. The second half is prepared in Orsay and will be measured by the AMS group in Aix en Provence.

[1] Clette et al., Space Sci. Rev. 286 (2014) 35
[2] Vonmoos et al., J. Geophys. Res. Space Phys. 111 (2006) 1
[3] Muscheler et al., Earth Planet. Sci. Lett. 219 (2004) 325

---

[1] *Geology, Lund University, Sweden*
[2] *EAWAG, Dübendorf*

A NEW ETH IN SITU COSMOGENIC ^{14}C EXTRACTION LINE

Building a second-generation extraction system

K. Hippe, M. Lupker[1], L. Wacker, C. Maden[2], H. Buseman[2], T.I. Eglinton[1], S. Ivy-Ochs, H.-A. Synal, S. Willett[1]

The main limitation for a widespread use of *in situ* ^{14}C analysis in Earth surface sciences is the technically demanding and time-consuming extraction procedure. A major challenge is the effective separation of the low abundances of *in situ* ^{14}C in quartz from the large amounts of contaminating atmospheric ^{14}C. ETH Zürich is one of only few places worldwide that have achieved *in situ* ^{14}C extraction. However, there is a strong need for further technical improvement as reliability and sample throughput represent major limitations of the current system.

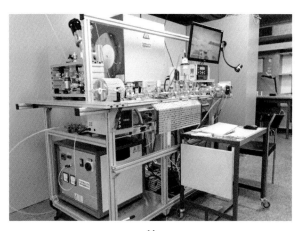

Fig. 1: *The new in situ ^{14}C extraction line.*

To overcome these technical challenges, in a combined effort of LIP with the groups of Biogeoscience, Geochemistry and Petrology, and Earth Surface, a new extraction system has been designed and assembled, and is now ready for testing (Fig.1). Main components of the improved system are:

- *In situ* ^{14}C extraction by high-temperature diffusion will be accomplished with a commercially available resistance furnace that promises highly reliable operation at the desired temperatures (1600°C).

- Automation of the system (Fig. 2) will enable performing large parts of the gas extraction overnight which is expected to at least double the overall sample throughput and reduce the attendance time of the researcher. It should also enhance the analytical reproducibility.

Fig. 2: *Pneumatic valves for automated gas extraction.*

- The gas purification process was simplified by removing unnecessary cold traps and reaction surfaces and by shortening extended gas flow paths that need long pump-down times.

- Instead of the present ultra-high vacuum extraction system helium carrier gas will be added to facilitate and speed up the gas transport and eliminate long pump-down times.

- "Dead" CO_2 carrier gas (containing no ^{14}C) will be added before extraction allowing to transfer a larger quantity of gas through the system and to reduce potential loss of sample gas through adsorption. This should result in a better reproducibility, especially for blanks and small samples.

---

[1] *Geology, ETH Zurich*
[2] *Geochemistry and Petrology, ETH Zurich*

SEDIMENT CONNECTIVITY DRIVES DENUDATION RATES

Measuring [10]Be in sediments from a heterogenic, alpine catchment

R. Grischott, F. Brardinoni[1], F. Kober[2], M. Christl

The formation of anomalously large alluvial fans in the Vinschgau (Val Venosta) in the Eastern Italian Alps has been widely debated. Based solely on the morphology, the landforms were thought to be the result of catastrophic massive slope failures [1]. However, analysis of historical regional data on debris-flow activity rather indicate that the alluvial fan area mirrors variable Holocene sediment transfer, which is a function of denudation of the upstream tributary basins. Denudation rates might be controlled by topographic metrics and rock type, debris availability and sediment transfer [2].

To quantify rates of denudation and sediment transfer, we extensively sampled sediments during three years for [10]Be analysis along the Strimm and Gadria Creek representing the source area of a large alluvial fan. Derived catchment-wide denudation rates (CWDRs) show a strongly heterogeneous pattern: Whereas in the Strimm basin the values increase downstream from 0.10 to 0.50 mm/yr, in the Gadria basin the values decrease downstream from 10.41 to 2.05 mm/yr (Fig. 1). CWDRs are routinely tested against topographic variables but exhibit poor correlation. By contrast, we find that the recently developed index of geomorphic connectivity (IC index), which integrates hillslope-to-channel sediment delivery and along-channel sediment conveyance [3], is a strikingly good metric for postglacial CWDRs (Fig. 1). Therefore, sediments from rapid headward erosion of the Gadria sub basin are efficiently transported via debris-flows to the final sink at the alluvial fan due to the overall high connectivity in the basin. Further research considering the volume and depth of the alluvial fan is pending.

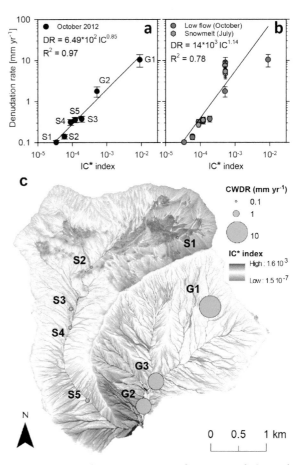

Fig. 1: *Denudation rate as a function of the IC* index in the Gadria and Strimm basins: (a) for samples collected in October 2012; and (b) the entire set of samples collected from October 2012 to July 2014, and stratified by season. (c) IC* index map with circle size proportional to the average (2012-2014) CWDR.*

[1] D. Jarman et al., Slope Tectonics (2011) 253

[2] F. Brardinoni et al., Geology 40 (2012) 455

[3] M. Cavalli et al., Geomorph. 188 (2013) 31

[1] *University of Bologna, Italy*

[2] *NAGRA, Wettingen*

CONSTRAINING EROSION RATES IN SEMI-ARID REGIONS

What is the limiting factor for soil chemical weathering and erosion?

V. Vanacker[1], J. Schoonejans[1], S. Opfergelt[1], Y. Ameijeiras-Mariño[1], M. Christl

Arid and semi-arid environments occupy around 37% of the land surface. The irregular rainfall regime with prolonged dry periods supports only sparse and patchy native vegetation cover. Sustained development of rain-fed agriculture is limited by soil and water resources [1].

This study focuses on the relationship between soil production, physical erosion, and chemical weathering. Study sites are located in the Southern Betic Cordillera (SE Spain), and selected across a spatial gradient in climatic and topographic conditions. The modern topographic expression of the Betic ranges reflects its tectonic history, with strong contrasts between the gently sloping hillsides of the Sierra de las Estancias, and the steep highly dissected landscape of the Sierra Cabrera.

Fig. 1: *Sampling of soil profiles in the Sierra Estancias (southeast Spain).*

Four catchments were selected to cover the range of denudation rates that were established for the Betic Cordillera [2]. In each catchment, two to three regolith profiles were sampled on

exposed ridgetops to avoid the complexities of soil forming processes associated with lateral transport of chemical fluids and soil particles along slope (Fig. 1). Total elemental composition of soil and rock samples was determined by ICP–AES (Thermo iCAP 6000 Series), and soil and sediment samples were processed for in-situ cosmogenic [10]Be analyses.

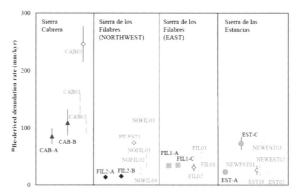

Fig. 2: *[10]Be-derived denudation rates of soils (colored symbols) and sediments (black symbols) in the Betic Cordillera.*

In the Southern Betic Cordillera, soil denudation rates are low, and range between 14 and 109 mm/kyr. Soil denudation rates are generally less than or equal to catchment-wide denudation rates measured at the outlet of small basins. Chemical weathering losses account for ~5 to 30 % of the total mass lost by denudation. Soil weathering increases (nonlinearly) with soil thickness and decreases with increasing surface denudation rates, consistent with kinetically limited weathering.

[1] V. Vanacker et al., Landscape Ecol. 29 (2014) 293

[2] N. Bellin et al., Earth Planet. Sci. Lett. 390 (2014) 19

[1] *Earth & Life Institute, University of Louvain, Belgium*

VEGETATION MODULATES DENUDATION RATES

^{10}Be measured in sediments from a strongly forested drainage basin

R. Grischott, F. Brardinoni[1], F. Kober[2], M. Christl

The topography of the Coast Mountains in British-Columbia (Canada) is strongly overprinted by repeated glaciations during the Pleistocene. Contemporary denudation rates based on sediment yield studies have shown to be still affected by so-called "paraglacial" sediment fluxes described as glacial sediments remobilized since deglaciation. For instance, sediment yields in the study area vary by one order of magnitude which raises the question what drives denudation and sediment export in this drainage basin [1].

To quantify rates of denudation and sediment transfer, we extensively sampled sediments for ^{10}Be analysis in the Capilano, Seymour and Lynn basins whose surfaces are mostly covered by glacial sediments. Derived catchment-wide denudation rates (CWDRs) show a rather homogeneous pattern with low rates of 0.2 to 0.3 mm/yr regardless of the basin size. Elevated CWDRs of 0.4 to 0.6 mm/yr were found for the small headwater catchments Sisters, Hesketh, East Cap (from west to east) and the larger Seymour basin (Fig. 1).

The low denudation rates of ~0.2 mm/yr are most likely due to the slope stabilization effect of the forested hillslopes. Despite this forest cover, the activity of debris-flows and landslides is rather high. However, field evidence has proven that these mass movements are shallow and thus the potential effect of "diluting" the ^{10}Be signal by low-dosed sediments from depth is minimal.

The two to three times higher rates found for the small headwater basins could be explained by fluctuating sediment yields due to episodic debris-flow activity (Fig. 2).

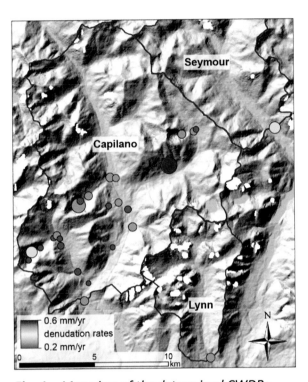

Fig. 1: *Map view of the determined CWDRs.*

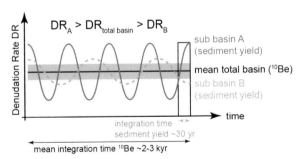

Fig. 2: *Schematic evolution of variable sediment yield cycles in different sub basins with respect to an assumed constant mean denudation rate of the total basin.*

[1]   F. Brardinoni et al., Geomorph. 49 (2003) 109

[1] *University of Bologna, Italy*
[2] *NAGRA, Wettingen*

THE PATTERN OF BACKGROUND EROSION RATE

Repeated measurements of ^{10}Be and landslide inventories

C.-Y. Chen[1], S.D. Willett[1], A.J. West[2], S. Dadson[3], N. Hovius[4], M. Christl, J.B.H. Shyu[6]

The southern Central Range of Taiwan is located at a tectonic transition zone and provides an opportunity to study the evolution of the mountain building at an early stage. To understand how this region evolves over time, it is essential to quantify erosion rates at different timescales. Here we apply a cosmogenic ^{10}Be study to characterize the pattern of denudation rates at millennial timescales.

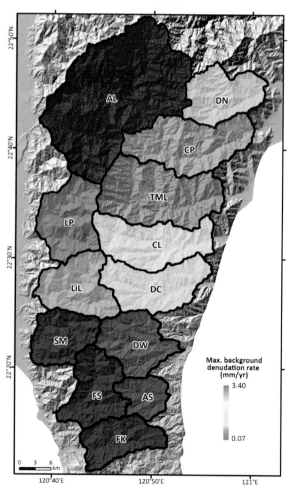

Fig. 1 *The pattern of maximum background denudation rate in the southern Central Range.*

The main challenge to this study is that about half of the basins in this region are highly impacted by landslides triggered by the typhoon Morakot in

2009. As deeply sourced landslide sediments can easily dilute in-situ ^{10}Be concentrations, caution must be taken in assessing the fraction of landslide sediments of each sample. In this study we use a simple sediment-mixing model parameterized by a time series of ^{10}Be concentrations derived from samples collected at different times and landslide inventories to constrain background rates and deliveries of landslide sediments.

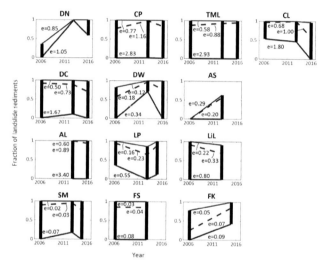

Fig. 2 *Ranges of background denudation rates and fractions of landslide sediments over time.*

[1] Geology, ETH Zurich
[2] Earth Sciences, University of Southern California, Los Angeles, CA 90089, USA
[3] Geography and Environment, University of Oxford, UK
[4] GFZ, Potsdam, Germany
[6] Geosciences, National Taiwan University, Taipei, Taiwan

TRACING LANDSCAPE EVOLUTION OF THE SILA MASSIF

Surface exposure dating of boulders using [10]Be

G. Raab[1], A. Ruppli[1], D. Brandova[1], F. Scarciglia[2], K. Norton[3], M. Christl, M. Egli[1]

Landscape surfaces and soils are known to evolve in complex, non-linear ways over thousands of years. As a result of changing environmental conditions over millennia, soil erosion and denudation processes also change substantially. Currently, our knowledge in this field is incomplete and very fragmented. Commonly, only averaged long-term erosion or denudation rates are determined for soils and landscapes. The fact that soil erosion processes are discontinuous over time is an aspect that is in most cases completely neglected. An evolutional approach is, thus, necessary to address such an issue. In the Sila Massif upland plateau (Calabria, Italy) boulders of different sizes are exhumed within an eroding landscape. To determine erosion rates and gain a better understanding of landscape evolution, surface exposure dating along vertical profiles of granitic boulders has been used (Fig. 1). Rock outcrops in erosive positions give additional indications about denudation rates.

Fig. 1: Spheroidal boulder outcrops in the Sila Massif. Their height (hm) can be used to calculate the minimum amount of eroded material [1].

[10]Be results of boulders have given a wide spectrum of ages, ranging from 10.7 ± 2.2 ka to 142.7 ± 28.7 ka. Ages were calculated using CRONUS 2.3 [2]. Although large boulders with heights of 2 to 6 m have been selected, some of them seemed to have been turned upside down or had a complex exhumation history that makes the interpretation of the data difficult. Some boulders, however, indicate a continuous vertical exhumation, although at differing rates over time. According to these first estimates, it seems that erosion rates (surface lowering rates) were highest before MIS 5b (> 90 ka; before the last glacial period) and during the Holocene (MIS 1). It seems that the cooler and drier conditions during the last glacial period reduced erosion rates whereas the moister conditions in the warmer periods enhanced surface lowering.

Further dating and measurements of cosmogenic nuclides in boulders and rock outcrops have to confirm this hypothesis. The approach seems, however, promising and delivers erosion and denudation rates during time segments of the Pleistocene and Holocene.

[1] F. Scarciglia, J. Soils Sediments 15 (2015) 1278

[2] G. Balco et al., Quat. Geochron. 3 (2008) 174

[1] University of Zurich
[2] University of Calabria, Arcavacata di Rende, Italy
[3] Victoria University, Wellington, New Zealand

ANTHROPOGENIC RADIONUCLIDES

5 years after Fukushima Daiichi nuclear accident

Investigation of environmental samples in Japan

Sampling sea and groundwaters around Fukushima

GRIFF: the expedition to the Fram Strait

Preliminary results of Transarc-II expedition

Anthropogenic ^{129}I in the North Atlantic

^{239}Pu and ^{236}U from eastern Tien Shan, China

A first transect of ^{236}U at the Equatorial Pacific

Radionuclides in drinking water reservoirs

Pu in snow on Mt. Zugspitze

Analysis of plutonium isotopes in aerosols

Chronometry model to date spent nuclear fuels

Analyses of proton-irradiated tantalum targets

5 YEARS AFTER FUKUSHIMA DAIICHI NUCLEAR ACCIDENT

Monitoring releases of long-lived radionuclides

N. Casacuberta, M. Christl, C. Vockenhuber, M. Castrillejo[1], P. Masqué[2], K.O. Buesseler[3]

The long-lived radionuclides ^{129}I, ^{236}U, ^{239}Pu and ^{240}Pu have been monitored in the coast off Fukushima 5 years after the nuclear accident, in Japan. Four different cruises took place during the years 2013-2015 within the remit of the EU FRAME project, and with the aim to re-evaluate the concentrations and distribution of the aforementioned radionuclides.

Results of ^{129}I concentrations in surface waters collected during these expeditions (Fig. 1) showed significantly high values (up to 700 x10^7 at·kg^{-1}) close to the Fukushima Daiichi Nuclear Power Plant (FDNPP).

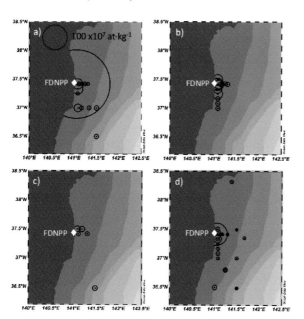

Fig. 1: Concentrations of ^{129}I measured during expeditions in: a) September 2013; b) May 2014; c) October 2014; and d) October 2015.

Although mean ^{129}I concentrations have decreased since 2011, our results evidenced the on-going releases of contaminated water to the marine environment. Yet, total releases of ^{129}I would not have been more than 0.1 kg added to the 1 kg released in 2011.

No ^{236}U concentrations exciding the global fallout signal were observed in seawater samples collected during the cruises in 2013 and 2014. This agrees with a previous study [1].

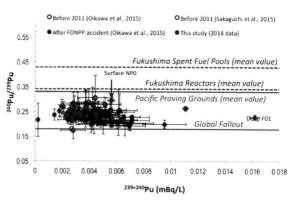

Fig. 2: $^{239+240}$Pu concentration and ^{240}Pu/^{239}Pu atom ratio from this study and others [1-2].

Pu-isotopes were only analyzed in samples collected during October 2014. The relationship between $^{239+240}$Pu concentrations and ^{240}Pu/^{239}Pu atom ratio gives an insight of the distribution of Pu-isotopes. Figure 2 shows that our values are within the range of atomic ratios expected from global fallout and Pacific Proving Ground (PPG). Only one sample collected close to the FDNPP (NP0) seems to point towards reactor grade Pu from the FDNPP.

[1] Sakaguchi et al, Geochem. J., 46 (2012) 355.

[2] Oikawa et al, J. Environ. Radioact., 142 (2015) 113.

---

[1] Universitat Autònoma Barcelona, Spain
[2] The University of Western Australia, Australia
[3] Woods Hole Oceanographic Institution, USA

INVESTIGATION OF ENVIRONMENTAL SAMPLES IN JAPAN

Origin of Pu and U in soil and litter samples near Fukushima Daiichi

S. Schneider[1], M. Christl, K. Shozugawa [2], G. Steinhauser[1], C. Walther[1]

During the accident of the power plant Fukushima Daiichi small amounts of refractory elements such as plutonium and uranium have been released. Previously investigated samples, collected in 2011 and 2013, showed a strong localization of plutonium [1]. These results fostered further investigations. New samples were taken in 2015 at nearly the same places as in 2013, which are shown in Fig. 1. At each sampling site a litter sample and a soil drill core, up to a depth of 15 cm, were collected. Each core sample was split into six equal parts. After chemical treatment, all samples were investigated using accelerator mass spectrometry. The isotopic ratios $^{240}Pu/^{239}Pu$ and $^{236}U/^{238}U$ were determined. These ratios are used as an indicator of the origin of the actinides, so it is possible to distinguish between global fallout and reactor material.

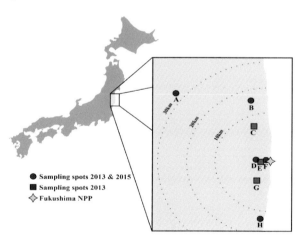

Fig. 1: *Map showing the sampling sites.*

In Fig. 2 the $^{240}Pu/^{239}Pu$ depth profiles are shown for all drill core samples. Most ratios agree with values of global fallout. Only Spot B shows higher values of 0.27 ± 0.02 and 0.29 ± 0.12, respectively. The first value clearly indicates reactor Pu, while for the second value global fallout cannot be completely ruled out due to its rather large uncertainty.

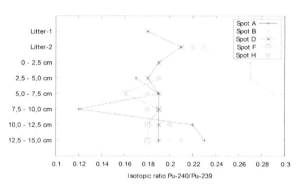

Fig. 2: Pu^{240}/Pu^{239} *ratio of the analyzed samples.*

In Fig. 3 the U^{236}/U^{238} ratio is shown for the analyzed samples. Most samples show values of a natural distribution. Only two litter samples show higher ratios in the range of 10^{-7}, which theoretically indicates an anthropogenic influence. However, such ratios were already measured in Japan before the accident.

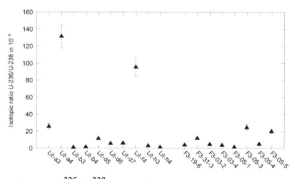

Fig. 3: U^{236}/U^{238} *ratio of the analyzed samples.*

[1] S. Schneider et al., Sci. Rep. 3, 2988; DOI:10.1038/srep02988 (2013)

[1] *Radioecology and Radiation Protection, University of Hannover, Germany*
[2] *Graduate School of Arts and Sciences, The University of Tokyo, Japan*

SAMPLING SEA AND GROUNDWATER NEAR FUKUSHIMA

Understanding radionuclide release

M. Castrillejo[1], P. Masqué[1,2], Jordi Garcia-Orellana[1], N. Casacuberta, M. Chirstl, C. Vockenhuber,
V. Sanial[3], K.O. Buesseler[3], M. Charette[3], S. Otosaka[4], H.-A. Synal

The Fukushima Dai-ichi nuclear power plant (FDNPP) has been releasing artificial radionuclides to the Pacific Ocean since the nuclear accident on March 11, 2011. Scientists from the Universitat Autònoma de Barcelona (UAB) and ETH-Zurich have been directly involved in the sampling, analysis and interpretation of several anthropogenic radionuclides from 2011 to 2016. The sampling in November 2016 was the result of the collaboration between UAB, ETH-Zurich, Woods Hole Oceanographic Institution (WHOI) and Japanese Institutions carried out within the EU COMET-FRAME project. This project aims at better understanding the releases from the FDNPP and their impact on marine sediments, water and biota.

Fig. 1: *Sampling of groundwater at a beach south of the FDNPP in Iwaki prefecture.*

Ground- and seawater were collected for the analysis of long-lived radionuclides (e.g. ^{129}I, ^{239}Pu and ^{240}Pu), cesium (^{137}Cs) and radium isotopes in a total of 50 samples per radionuclide. The analysis of ^{129}I and the Pu isotopes will be done using the 0.5 MV TANDY Accelerator Mass Spectrometry facility at ETH-Zurich. The results of ^{129}I will be combined with data on Ra isotopes and ^{137}Cs from WHOI to study the radionuclide desorption from beach sediments to the seawater column. Ra isotopes will be used as 'chronological tracers' to provide an estimate of the radionuclide flux associated to the discharge of submarine ground waters.

Fig. 2: *Japanese research vessel Sinshei Maru ready to sail before the November 2016 cruise.*

Data from groundwaters will be interpreted in a broader context including the monitoring of the seawater column within <30 km off the FDNPP. The results should also allow characterizing the different sources of radioactivity using radionuclide ratios (e.g. ^{129}I/^{137}Cs). A summary of the ^{129}I and Pu isotopes data from 2013 to 2015 is provided on page 80 in this issue.

---

[1] *Institut de Ciència i Tecnologia Ambientals & Departament de Física, Universitat Autònoma de Barcelona), Spain*
[2] *Edith Cowan University, Western Australia*
[3] *Woods Hole Oceanographic Institution, MA USA*
[4] *Japan Atomic Energy Agency, Japan*

GRIFF: EXPEDITION TO THE FRAM STRAIT

Collecting seawater samples for ^{236}U and ^{129}I

N. Casacuberta, M. Christl, C. Vockenhuber, M. Rutgers van-der-Loeff[1], H.-A. Synal

The 18th of July was an important date: the GRIFF expedition started! 48 scientists from 10 different countries, together with 50 crew members, left Tromsø onboard one of the most supreme research vessels: Polarstern (Fig. 1). We left a grey and rainy day behind us, but excitement was shinning in all our faces while sailing through the fjords to reach the open ocean.

Fig. 1: *R/V Polarstern during GRIFF expedition; a view from the coast of Greenland.*

During the 7 weeks of the GRIFF expedition we took more than 300 seawater samples for the analysis of ^{129}I and ^{236}U, covering the entire section from Svalbard to the East coast of Greenland.

Even before getting any results, our eyes witnessed the oceanic differences between Svalbard and Greenland. In the Eastern part of the transect (close to Svalbard) and while crossing the West Spitzbergen Current we had a summer feeling due to the relatively warm Atlantic water flowing northward to the Arctic Ocean. But this feeling changed in a mere 150 km to the West, where the whole landscape transformed to a winter environment while crossing the East Greenland Current (Fig. 2). This current carries polar water covered by the ice formed in the Arctic Ocean, back to the Atlantic Ocean. Here we also find presence of the so-

called "dirty" ice that is brown to black from its heavy load of mud. This muddy ice crossed the Arctic Ocean in the Transpolar Drift, the ice current that Nansen used in his attempt to reach the North Pole. The presence of this current was confirmed when we recovered a buoy that Polarstern had deployed in the central Arctic just a year ago.

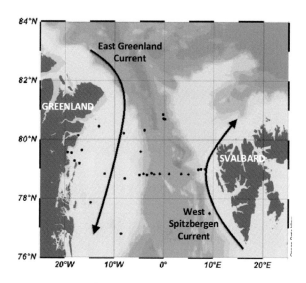

Fig. 2: *Sampling sites of the GRIFF expedition.*

Results of ^{129}I and ^{236}U, together with other parameters collected by the GEOTRACES team will help understanding biogeochemical cycles of trace elements and their isotopes in the Arctic Ocean. We expect to find a much stronger signal of both ^{129}I and ^{236}U coming from Sellafield and La Hague in the West Spitzbergen Current [1], compared to the outflowing East Greenland Current, where waters are older and the reprocessing signal is much more diluted.

[1] Casacuberta et al., Earth Planet. Sci. Lett. 440 (2016) 127

---

[1] *Alfred Wegener Institute, Germany*

PRELIMINARY RESULTS OF TRANSARC-II EXPEDITION

^{236}U and ^{129}I: from the Barents Sea to the North Pole

N. Casacuberta, M. Christl, C. Vockenhuber, M. Rutgers van-der-Loeff[1], H.-A. Synal

The GEOTRACES section (GN04, TransArc-II) on board the German R/V Polarstern covered a full transect in the Arctic Ocean from the Barents Sea to the Makarov Basin, crossing the North Pole. The distributions of ^{129}I concentrations and ^{236}U/^{238}U atomic ratios obtained from more than 300 seawater samples are consistent with different water masses (Fig. 1).

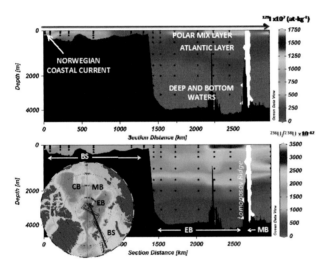

Fig. 1: *^{129}I concentrations and ^{236}U/^{238}U atomic ratio of the transect from Barents Sea (BS) to the Makarov Basin (MB).*

Atlantic Waters flowing into the Arctic Ocean carry the signal of the two European nuclear reprocessing plants of Sellafield (SF) and La Hague (LH), with the highest ^{129}I concentrations (> 1500x10^7 at·kg^{-1}) and ^{236}U/^{238}U ratios (> 3000x10^{-12}) in the Norwegian Shelf, where the Norwegian Coastal Current penetrates into the Barents Sea. Lowest ^{129}I concentrations (< 5x10^7 at·kg^{-1}) and ^{236}U/^{238}U ratios (< 10x10^{-12}) were observed in the deep and bottom waters of the Makarov Basin, proving the long-term isolation of these waters. The combination of ^{129}I/^{236}U and ^{236}U/^{238}U atomic ratios can be used to identify sources of artificial radionuclides to the Arctic Ocean [1].

This dataset confirms that global fallout and reprocessing plants are the main suppliers of ^{129}I and ^{236}U to the Arctic Ocean, while Siberian rivers would be minor contributors. Different from previous assumptions, results show that the Barents Sea Branch Water (BSBW) has a higher ^{129}I/^{236}U atom ratio than expected from the mixed signal of SF and LH in the previous 5 years (Fig. 2).

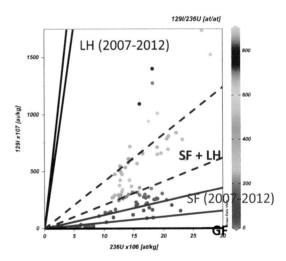

Fig. 2: *^{129}I and ^{236}U concentrations plotted over the expected ^{129}I/^{236}U atomic ratios from SF, LH and the mixture of both (SF+LH).*

This high ^{129}I/^{236}U ratio suggests that the contribution of LH relative to SF is larger than expected in this water mass. Together with the data from other GEOTRACES cruises, a synoptic distribution of ^{129}I and ^{236}U in the Arctic Ocean will be available soon, which will help understanding major water circulation patterns in the Arctic Ocean.

[1] Casacuberta et al., Earth Planet. Sci. Lett., 440 (2016) 127

---

[1] *Alfred Wegener Institute, AWI, Germany*

ANTHROPOGENIC ^{129}I IN THE NORTH ATLANTIC

^{129}I along the zonal GEOVIDE transect

M. Castrillejo[1], P. Masqué[1], Jordi Garcia-Orellana[1], N. Casacuberta, C. Vockenhuber, M. Chirstl

The distribution of ^{129}I along the zonal GEOVIDE transect (black solid line, Fig. 1) was investigated in spring 2014. In this region, the presence of ^{129}I is largely due to releases from Sellafield and La Hague reprocessing plants, and to a lesser extent, to global fallout from the atmospheric nuclear bomb tests carried out from 1945 to 1980. In this study data on ^{129}I from over 150 seawater samples are interpreted in the context of the above sources and the regional water circulation.

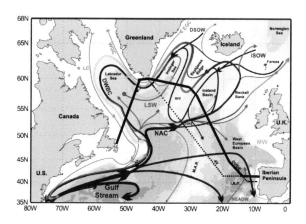

Fig. 1: *Schematic North Atlantic circulation [1].*

The concentrations of ^{129}I (Fig. 2) range between <1×10^{7} and 20×10^{7} at·kg^{-1} east of Reykjanes Ridge. To the west, the concentrations of ^{129}I are much larger and range between 10×10^{7} at·kg^{-1} and >250×10^{7} at·kg^{-1}.

The distribution of ^{129}I is consistent with the presence of different water masses carrying varying amounts of ^{129}I from reprocessing and global fallout. Radionuclide-poor waters of Antarctic origin fill the depths below 2000 m east of 25 ºW (purple color in Figure 2). These waters appear to carry small amounts of ^{129}I (<5×10^{7} at·kg^{-1}) from global fallout or show no anthropogenic influence. On the contrary, waters in the Irminger and Labrador Seas (Fig. 1) record concentrations of ^{129}I up to 2 orders of

magnitude larger indicating the presence of releases from Sellafield and La Hague (red color in Fig. 2). These radionuclide-rich waters flow from the Arctic into the North Atlantic through the Iceland-Scotland and the Iceland-Greenland passages.

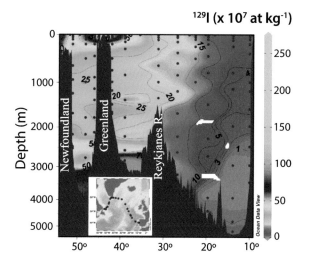

Fig. 2: *Distribution of ^{129}I from the Iberian Peninsula (right) to Newfoundland (left).*

The high ^{129}I concentrations in the deep Irminger and Labrador Seas show among other features, the formation and sinking of radionuclide-rich waters. In future work, ^{236}U will be measured and combined with ^{129}I in a dual tracer approach (^{129}I/^{236}U - ^{236}U/^{238}U) to estimate tracer ages and ventilation rates as done for the Northeast Atlantic [2].

[1] García-Ibañez et al., Prog. In Ocea., 135 (2015) 18-36

[2] Christl et al., J. Geophys. Res. Oc., (2015)

[1] *Institut de Ciència i Tecnologia Ambientals & Dept. de Física, Universitat Autònoma de Barcelona, Spain*

^{239}Pu AND ^{236}U FROM EASTERN TIEN SHAN, CHINA

Total flux of actinide deposition at the Miaoergou glacier

C. Wang[1], H.W. Gäggeler[2], S. Hou[1], H. Pang[1], Y. Liu[3], M. Christl, H.-A. Synal

Anthropogenic radioactive elements have been released by human activities. ^{239}Pu and ^{236}U deserve special attention due to their high radiological toxicity and long persistence in the environment. Since the first nuclear weapons test in 1945, ^{239}Pu and ^{236}U have been dispersed into the environment.

In 2005, a 57.6 m ice core to bedrock was recovered from a dome on the Miaoergou glacier, eastern Tien Shan, central Asia (43°03'19"N, 94°19'21"E, 4512 m a.s.l.). The low borehole temperature at the drilling site (-7.2 °C at 10 m depth and -8.2 °C at the bottom, respectively) is beneficial for the preservation of ice core records [1]. The age-depth relationship of the upper 42.4 m w. e. of the core was established by the combination of annual layer counting, ^{210}Pb dating and detection of the nuclear test time marker (beta activity) [1, 2, 3]. The core segments from 8 to 15 meter depth corresponding to the period between about 1940 and 1970 [2] were selected for AMS analysis with the compact low energy system Tandy at the Laboratory of Ion Beam Physics, ETH Zürich [4].

The total deposition of ^{239}Pu integrated over the nuclear weapons testing period (NWT) amounts to 1.55 x 10^9 atoms·cm^{-2} (Tab. 1). This value is higher compared to values reported for glaciers from the European Alps with 0.9x10^9 atoms·cm^{-2} at Col du Dome, a site close to Mont Blanc in France and 0.7x10^9 atoms·cm^{-2} at Colle Gnifetti, Swiss Alps [5] but lower than 3.6x10^9 atoms·cm^{-2} obtained from an analysis of an ice core at Belukha, Altai (Russia) [6]. Other literature values are 1.7x10^9 measured in a Greenland ice core [7] and 0.54x10^9 at the Agassiz ice cap [8]. In a recent publication [9] a comparison was made between ^{239}Pu concentration in Arctic and Antarctic samples, indicating that they differ by about a factor of three (Northern hemisphere to Southern hemisphere ratio). The maximum

value for ^{239}Pu from several Arctic sites (about 5 mBq/kg) agrees well with our maximum value of 5.5 mBq/kg. A total fallout of ^{236}U from NTW with the value of 3.5x10^8 atoms·cm^{-2} agrees rather well with 1.63x10^8 atoms·cm^{-2} [10] from the Arctic site Svalbard 79.83°N based on the expected trend of deposition rate with latitude.

Site	Latitude	Longitude	Total Flux (at·cm^{-2})
Miaoergou	43°03'19"N	94°19'21"E	1.55 x 10^9
Col du Dome	/	/	0.9 x 10^9
Colle Gnifetti	45°55'50.4"N	07°52'33.5"E	0.7 x 10^9
Belukha	49°48'26"N	86°34'43"E	3.6 x 10^9
Agassiz	/	/	0.54 x 10^9
Greenland	65°11'N	43°50'W	1.7 x 10^9

Tab. 1: Total fluxes of ^{239}Pu from different ice core records.

[1] Y. Liu et al., J. Geophys. Res. 116 (2011) D12307
[2] C. Wang et al., Ann. Glaciol. 55 (2014) 66
[3] C. Wang et al., Ann. Glaciol. 57 (2016) 71
[4] M. Christl et al., NIM B 294 (2013) 29
[5] J. Gabrieli et al., Atmos. Environ. 45 (2011) 587
[6] S. Olivier et al., Env. Sci. Tech. 38 (2004) 6507
[7] M. Koide et al., EPSL 72 (1982) 1
[8] A. Kudo et al., Water Sci. Tech. 42 (2000) 163
[9] M. Arienzo et al., Env. Sci. Tech. 50 (2016) 7066
[10] C. Wendel et al., Sci. Tot. Env. 461 (2013) 734

[1] Nanjing University, China
[2] Paul Scherrer Institut, Villigen
[3] Northwest Institute of Eco-Environment and resources, Chinese Academy of Sciences, China

A FIRST TRANSECT OF ^{236}U AT THE EQUATORIAL PACIFIC

Profiles from the GEOTRACES East Pacific Zonal Transect (EPTZ)

M. Villa-Alfageme[1], E. Chamizo[2], M. López-Lora[2], N. Casacuberta, T. Kenna[3], J.M. López[1], M. Christl

The U.S. GEOTRACES East Pacific Zonal Transect (EPZT), from Peru to Tahiti, was occupied in 2013. Four complete depth profiles, approximately 100 samples, were processed for ^{236}U measurement (i.e. ^{236}U/^{238}U atomic ratios (AR) and ^{236}U concentrations). Actinides were co-precipitated at Lamont-Doherty Earth Observatory (LDEO) using 4 L samples. ^{236}U radiochemical treatment was performed at CNA [1]. Samples above 400 were measured on the 1 MV AMS CNA facility, (^{236}U/^{238}U abundance sensitivity of about 9×10^{-11}). AR in deeper samples were measured at the 600 kV AMS ETH Tandy facility in Zürich.

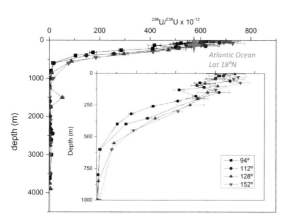

Fig. 1: a) ^{236}U/^{238}U ($\cdot 10^{-12}$) measured at the East Pacific Zonal Transect b) zoomed above 1000 m. The coordinates in the legend correspond to the longitude.

We find the following relevant results: (a) the concentration profiles are characteristic of a conservative element with an atmospheric input at surface; (b) surface ^{236}U/^{238}U AR are consistent with their latitude (Fig. 1) [2]; (c) inventories range from 1.45×10^{12} to 2×10^{12} at·m^{-2}, in agreement with values in the Southern Hemisphere exclusively affected by global fall-out; and d) ^{129}I/^{236}U ratios are ~1.2 at at^{-1}, typical of a global fall-out input. Furthermore, St. 18 (112°W), corresponds to a hydrothermal

plume (HP) originating from the southern East Pacific Rise, enhancing concentration of specific elements. However the variations of ^{236}U here are not statistically significant, according to their uncertainties.

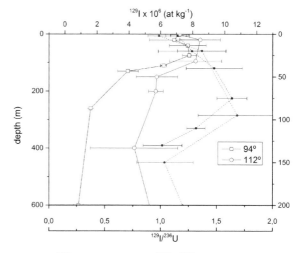

Fig. 2: ^{129}I (empty) and ^{129}I/^{236}U (solid) profiles measured at 94° and 112°.

^{236}U concentrations measured in this work were the lowest found so far using ETH Tandy (at the 10^{-12} level). We conclude that there are no detectable ^{236}U released from local sources and concentrations at EPZT corresponds to global fall-out.

[1] M. López-Lora et al., Talanta. Submitted.
[2] N. Casacuberta et al., Geochim. Cosmoschim. Acta 133 (2014) 34

^{1} *Applied Physics Dpt. Universidad de Sevilla, Spain*
^{2} *Centro Nacional de Aceleradores (CNA). Universidad de Sevilla, Spain*
^{3} *Lamont-Doherty Earth Observatory, Columbia University, USA*

RADIONUCLIDES IN DRINKING WATER RESERVOIRES

Sensitivity of reservoirs to input of man-made radionuclides

B. Riebe[1], S. Bister[1], A.A.A. Osman[1], A. Daraoui[1], C. Walther[1], C. Vockenhuber, H.-A. Synal

Water as an essential resource for life is exposed to different environmental pollutants originating from human activities, including radionuclides. In a current project we are aiming to assess the sensitivity of an unconfined aquifer in Northern Germany (Fuhrberger Feld), which is used as a drinking water reservoir, with regard to introduction and accumulation of radionuclides, e.g. ^{3}H, ^{14}C, ^{90}Sr, ^{137}Cs and ^{129}I.

Ground water samples are drawn from multilevel-wells at different depth (4 m, 14 m) aligned in the direction of ground water flow. Additionally, adjacent rivers and ponds were sampled (Fig. 1). A detailed description of sample preparation is given elsewhere [1]. For analysis of ^{129}I accelerator mass spectrometry (AMS) is used, ^{127}I concentrations are determined by inductively coupled plasma mass spectrometry (ICP-MS).

Fig. 1: *Drawing surface water samples from a pond in the 'Fuhrberger Feld'.*

Results of the analyses reveal that ^{129}I activity concentrations in water samples from Fuhrberger Feld (FF) are about one order of magnitude higher than those from water samples from other aquifers from Lower Saxony (LS) (Fig. 2). For FF activity concentrations of ^{129}I

in the ground water samples vary between 1.4×10^{-7} and 5.2×10^{-7} Bq kg^{-1}, and therefore lie within the range of values determined for surface water samples from LS. In comparison, ^{129}I activity concentrations measured for samples of confined aquifers from the same region are as low as 8.5×10^{-9} to 1.0×10^{-7} [2].

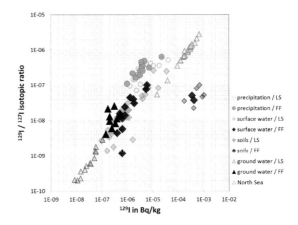

Fig. 2: *$^{129}I/^{127}I$ isotopic ratios versus ^{129}I for water samples from Fuhrberger Feld (FF) in comparison to other environmental compartments in Lower Saxony (LS), and North Sea water [1].*

The same is true for the $^{129}I/^{127}I$ ratio. Values of FF ground water samples (4.1×10^{-9} and 1.5×10^{-8}) are in the range of surface water samples from the same sampling area, whereas samples from confined aquifers show values, which are one order of magnitude lower. This indicates that ^{129}I from the atmosphere has already reached the Fuhrberger Feld aquifer.

[1]   R. Michel et al., Sci. Tot. Environ. 419 (2012) 151

[2]   A.A.A. Osman et al., J. Environ. Rad. 165 (2016) 243

[1] *Radioecology and Radiation Protection, Leibniz University Hannover, Germany*

PU IN SNOW ON MT. ZUGSPITZE

Determination of Pu isotopes precipitated with snow on Mt. Zugspitze

K. Gückel[1], M. Christl

For the forecast of radiological consequences of an accidental release of actinides into the environment, it is important to know the long-term behavior of those elements. Since the first explosion of a nuclear weapon (Trinity test) in 1945, Pu isotopes are being released into the environment through e.g. nuclear weapon tests, accidents of nuclear facilities, satellite and plane crashes, and discharge from reprocessing plants. It is assumed that around $1.4 \cdot 10^{16}$ Bq of $^{239+240}$Pu were released into the environment ([1] and references therein).

However limited data of the Pu amount wet-deposited in Bavaria and the alpine area exists [2-4] and details on the transport of actinides from snow areas into the hydrosphere are still not entirely understood.

In this study, we focus on the analysis of quantity and atomic ratios of Pu isotopes in the global fallout on Mt. Zugspitze. For the attogram level (10^{-18} gram) of Pu a chemical separation procedure was improved using extraction chromatography. The compact Tandy AMS was used for the measurement of isolated Pu.

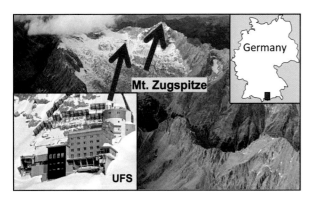

Fig. 1: *Study site at UFS on Mt. Zugspitze*

Sampling (60-140 kg of snow) was done at the Environmental Research Station Schneeferner-haus (UFS) on Mt. Zugspitze (Fig. 1). The UFS is located at the south slope of Mt. Zugspitze,

which is part of the Wettersteingebirge in the south of Germany. With an elevation of 2650 m a.s.l. the station represents a high alpine site.

Undisturbed fresh fallen snow was collected in eight 40 L wide-necked barrels and acidified with 65 % HNO_3. The snow melt was pre-concentrated by evaporation with a large capacity evaporator in the radiochemistry lab of the institute of radiation protection, before the chemical sepration and AMS sample preparation.

The concentrations in the fresh fallen snow range from (75 ± 14) to (2800 ± 84) ag/kg of ^{239}Pu, from (21 ± 5) to (600 ± 21) ag/kg of ^{240}Pu, and from (1.5 ± 1.4) to (12 ± 7) ag/kg of ^{241}Pu. The ^{240}Pu/^{239}Pu atomic ratios are comparable to the global fallout in middle Europe but input of resuspended particles of Sahara dust is considered.

The results give a first impression of the amount of Pu isotopes deposited on the snow area at Mt. Zugspitze. Analysis of additional snow samples from future winters and further studies of Pu in the atmosphere will help to understand the behavior of Pu in the global fallout-snow-hydrosphere system.

[1] T.Bisinger, Südwestdeutscher Verlag für Hochschulschriften, 2010
[2] J. Gabrieli, et al., Atmos. Environ., 45 (3) (2011) 587
[3] G. Rosner, R. Winkler, Total Environ, 273 (2001) 11
[4] Warneke, et al., Earth Planet. Sci. Lett., 203 (2002) 1047

[1] *Helmholtz Zentrum München, Deutsches Forschungszentrum für Gesundheit und Umwelt (GmbH)*

ANALYSIS OF PLUTONIUM ISOTOPES IN AEROSOLS

Method development for determination of ^{239,240}Pu in air filters

X. Dai[1], M. Luo[1], M. Christl, J. Ouyang[2], D. Xu[2]

During the years of the large-scale nuclear weapon tests, bomb-produced plutonium was directly injected into the stratosphere, reached the troposphere especially in late spring when rising hot air provokes the descent of cold air masses from the lower stratosphere, and finally precipitated to the ground according to the meteorological conditions [1]. These typical spring maxima of Pu in ground-level air caused by the stratosphere-troposphere exchange were observed till the 1990s. Lately, the variations of Pu isotopes in the air have been governed mostly by their resuspension from soil, indicating that most of atmospheric Pu from the earlier nuclear weapon tests had already deposited on land and ocean surface. At present, the concentrations of Pu isotopes in air have dropped by ~3 orders of magnitudes (down to the nBq/m^3 levels for ^{239,240}Pu), which have become very difficult to measure. To continue the research on Pu distribution and transport behavior in the atmosphere, more sensitive methods for Pu in aerosol samples, particularly by AMS, are needed.

In this work, a radiochemical separation method for the analysis of Pu by AMS in aerosol samples was developed (Fig. 1). Lithium metaborate fusion technique was used for complete sample dissolution. The fusion button was dissolved with mixed HCl/HNO$_3$ acid, and the silicates were removed by PEG (polyethylene glycol) flocculation. A hydrous titanium oxide (HTiO) co-precipitation step was then conducted to remove the soluble salts, followed by the Pu purification using the TEVA column. After elution from the resin, the Pu was subsequently co-precipitated with mixed Ti/Fe hydroxides, dried, and pressed to the AMS target with the addition of niobium powder. Finally, the Pu isotopes were analyzed using compact AMS Tandy at ETH.

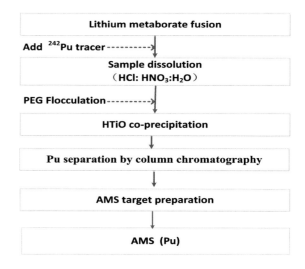

Fig. 1: *Flow diagram of the radiochemical separation procedure for Pu in aerosol samples.*

Four aerosol samples collected in Beijing in May and July 2016, along with two procedural blanks, were analyzed for initial testing of the method performance. The volume of filtered air varied from 500 to 1500 m^3 with the aerosol particulates ranged from 50 to 250 mg. The method was found to be sufficiently sensitive to detect <0.8 ag/m^3 for ^{239}Pu (<1.7 nBq/m^3) and <0.3 ag/m^3 for ^{240}Pu (<1.8 nBq/m^3), respectively. Preliminary results also showed the isotopic ratio of 0.16±0.14 for ^{240}Pu/^{239}Pu (at/at) on the aerosol samples, which is consistent with the Pu ratio for global fallout.

[1] K. Hirose and P. P. Povinec, Scientific Reports, 5 (2015) 15707

[1] *China Institute for Radiation Protection*
[2] *Institute of High Energy Physics, Chinese Academy of Sciences*

CHRONOMETRY MODEL TO DATE SPENT NUCLEAR FUELS

First conceptual results using Cm isotopic ratios

S. Burrell[1], M. Christl, A. Gagné[1], N. Guerin[1], M. Totland[1]

Illicit trafficking of nuclear material is a serious threat in the hands of criminal and terrorist organizations because the material could be used to prepare radiological dispersive devices (RDD) [1]. When an illicit nuclear material is found, nuclear forensics analyses are conducted on the radioactive sample to determine its chemical/radiological composition, physical structure, and age since production or use. The information obtained is used to determine where and when the illicit material was produced, purified or enriched, irradiated, and possibly taken from regulatory control.

Currently, the main technique used to calculate when a nuclear material was prepared is by using radioactive in-growth equations: a parent isotope decays into a daughter isotope at a constant rate (e.g. expressed as radiological half-life). By measuring the activities of the parent and daughter isotope (for example $^{241}Am/^{241}Pu$) it is possible to find out when the parent was isolated. This strategy is appropriate for materials where a parent isotope has been isolated (such as military grade Pu), but is less accurate for irradiated fuel samples. In spent nuclear fuels, a large number of isotopes are formed by fission and neutron capture. The amount of radionuclide daughters generated by the parent cannot be differentiated from the amount of radionuclide daughter initially present in the sample or generated via different nuclear pathways; thus, new more accurate strategies need to be developed to date irradiated materials.

This work investigates the possibility of age-dating spent nuclear fuel samples utilizing the fact that ^{245}Cm to ^{249}Cm isotopes can only be formed by neutron capture from ^{244}Cm (Fig. 1).

This year, code data (calculated quantities of isotopes using mathematical models) has been used to demonstrate that there is a relation between some Cm isotopic ratios (Fig. 2) for materials with different irradiation times.

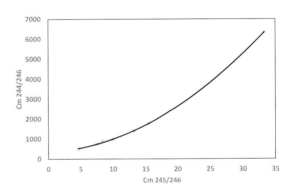

Fig. 2: $^{244}Cm/^{246}Cm$ *as a function of* $^{245}Cm/^{246}Cm$ *ratio calculated from code data.*

Verification of the code data relation has started. A number of available irradiated fuel materials were identified and work has started to chemically isolate Cm from highly radioactive solutions. The objective for next year is to compare the curves obtained from code data to real sample measurements. AMS is the ideal technique to determine Cm isotopic ratios because of its high sensitivity.

[1]   IAEA, Combating illicit trafficking in nuclear and other radioactive material, reference manual, Vienna 2007

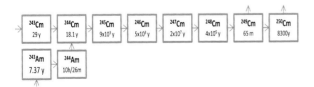

Fig. 1: *Neutron absorption chart.*

[1] *Canadian Nuclear Laboratories, Chalk River, Canada*

ANALYSES OF PROTON-IRRADIATED TANTALUM TARGETS

Cross section measurements of ^{129}I

Z. Talip[1], C. Vockenhuber, J.-Ch. David[2], D. Schumann[1]

Knowledge about the residual nuclide production in target materials is essential for the safety of spallation neutron sources and accelerator driven systems (ADS). In addition, it is important for evaluation and improvement of computer simulations for many different target-energy combinations. In this context we determined the production cross sections of the long-lived β-emitting radionuclide ^{129}I ($T_{1/2}$: 15.7 My) in p-irradiated Ta targets, using the Tandy AMS facility.

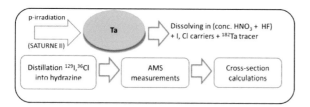

Fig. 1: *Schematic diagram for the experimental cross section determination of ^{36}Cl and ^{129}I from p-irradiated Ta targets.*

Fig. 1 shows the experimental procedure: Ta targets (typical thickness 125 mg/cm^2, 15 mm diameter, 99.99% purity) were irradiated with 240 to 2600 MeV protons between 1993-97 with a special stacked foil technique at the SATURNE II synchrotron of the Laboratorie National Saturne (LNS) at Saclay [1]. Then the Ta targets, together with I carrier (10 mg NaI from Woodward Corporation, USA), Cl carrier (10 mg NaCl from Riedel-de-Haen) and ^{182}Ta as tracers, were dissolved in 10 mL 10 M HNO$_3$ and 4 mL conc. HF in a PTFE (Polytetrafluoro-ethylene) two neck-flask in an N$_2$ atmosphere at 100 °C. I and Cl were distilled into an aqueous hydrazine solution (1:1, total 7 mL). The separation of I and Cl was performed similar to [2]. The ^{129}I/^{127}I ratios were measured at the Tandy AMS facility and were ranging from 0.2×10^{-12} to 0.54×10^{-12}. From 11 samples only 6 showed significant values above the chemistry blank (0.2×10^{-12}, Fig.

2a). Thus the final errors of the cross section after the blank correction are relatively large (Fig. 2b). For comparison the theoretical cross section calculations using the combination of Liège intranuclear cascade (INCL 4.6) and de-exitation phase (ABLA07) codes [3] are also shown.

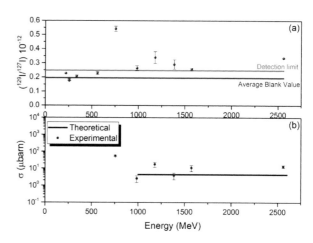

Fig. 2: *Results of the AMS measurements without blank correction (a), experimental and theoretical cross section results for ^{129}I in proton-irradiated Ta (b).*

Main uncertainty components considered for the cross section calculations are due to the flux density (6%) and AMS measurements (~8-42%). Further AMS measurements will be performed to measure production cross sections of the long-lived β-emitting radionuclide, ^{36}Cl ($T_{1/2}$: 301 ky) in proton-irradiated Ta targets.

[1] M. Gloris, PhD thesis, Hannover (1998)
[2] B. Rotzler et al., Radiochim. Acta 103 (2015) 745
[3] D. Mancusi et al., Phys. Rev. 90 (2014) 054602

^{1} PSI, Villigen
^{2} Irfu, CEA, Universite Paris-Saclay, France

MATERIALS SCIENCES

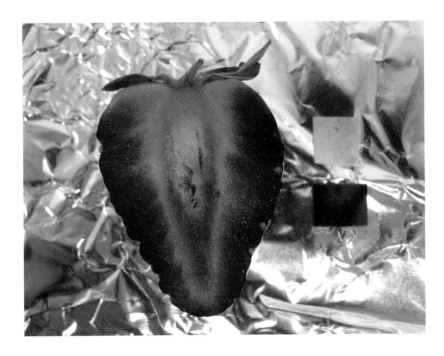

MeV-SIMS: Nuclear versus electronic stopping

MeV-SIMS imaging with CHIMP

Modified confocal DC magnetron sputtering

Porous metal films

Mechanical properties of thin film stacks

High boron-content graphitic carbon

Total IBA

Selective grain growth by ion-irradiation

Production cross-section of $^{42}Ca(d, n)^{43}Sc$

MEV-SIMS: NUCLEAR VERSUS ELECTRONIC STOPPING

Reducing secondary ion fragmentation by swift heavy ion sputtering

K.-U. Miltenberger, M. Schulte-Borchers, A.M. Müller, M. Döbeli, H.-A. Synal

Conventional secondary ion mass spectrometry (SIMS) relies on the interaction of the primary ion beam with target nuclei to induce sputtering of secondary particles from the sample surface by nuclear energy loss. Generally, this results in very high fragmentation of emitted molecules. By using primary ions with high electronic energy loss, fragmentation can be considerably reduced which significantly facilitates the identification of large molecules present in the material. The contribution of each energy loss mechanism depends strongly on the primary ion velocity and the atomic number Z of the primary ions. For slow velocities and high Z nuclear stopping dominates, while electronic stopping is increasingly important for higher velocities and lower Z (Fig. 1).

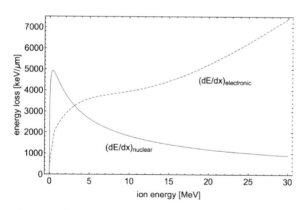

Fig. 1: *Electronic (dark) and nuclear (light) energy loss of Au ions in silicon nitride as a function of ion energy as calculated by SRIM [1].*

While the underlying processes are not understood in detail, experiments have demonstrated a significant increase in molecular secondary ion yields [2] when moving to swift, heavy primary ions (MeV-SIMS). This could be confirmed using the new MeV-SIMS setup at ETH Zurich [3].

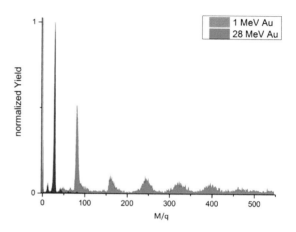

Fig. 2: *Positive secondary ion mass spectrum of a silicone sample with a 1 MeV (red) and 28 MeV (blue) primary Au ion beam [3].*

Measurements were done on a silicone-coated Si wafer with primary Au ion beams of 1 MeV and 28 MeV respectively (Fig. 2). The reduced fragmentation of secondary ions sputtered with the 28 MeV Au beam is apparent. While only ions with a mass of up to 100 u were detected in the 1 MeV Au spectrum and just the peak from a single PDMS fragment $[C_2H_6OSi]$ is clearly visible, the 28 MeV Au spectrum also contains higher oligomer $[C_2H_6OSi]_n$ peaks and the detected mass range is extended to masses of several hundred u.

In a next step the wide range of primary ion beams available from the 6 MV EN-Tandem accelerator will be used to study secondary ion yields in more detail.

[1]   J.F. Ziegler, SRIM 2013

[2]   P. Hakansson et al., Nucl. Instr. & Meth. 198 (1982) 43

[3]   M. Schulte-Borchers, Diss. ETH No. 23600 (2016)

MEV-SIMS IMAGING WITH CHIMP

Molecular imaging with a capillary microprobe

K.-U. Miltenberger, M. Schulte-Borchers, A.M. Müller, M. Döbeli, H.-A. Synal

The new MeV-SIMS microprobe setup CHIMP (Capillary Heavy Ion MeV-SIMS Probe) developed and built at the ETH Zürich 6 MV EN tandem accelerator facility [1] enables imaging of secondary molecular ions sputtered by primary MeV heavy ions. The primary beam is collimated to a diameter of 1 – 10 µm (divergence < 0.3 mrad) by a tapered glass capillary . Due to the high sputter yields of MeV ions a primary beam of a few fA is sufficient to produce secondary particle count rates of the order of 10 kHz. Positive secondary ions are detected in a Time-of-Flight (ToF) mass spectrometer with the start signal provided by a secondary electron detector.

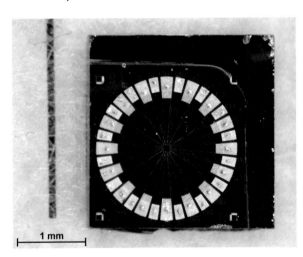

Fig. 1: *GaAs chip with alternating Al and Au contact pads arranged in a circular pattern.*

To enable imaging the sample is mounted on a piezo driven xy raster stage orthogonal to the primary ion beam, which can be retracted completely out of the beam path for beam current measurements and sample changes. Data acquisition is handled by a CAEN 4-channel digitizer recording all detected events as well as the sample position into separate files, which can be analyzed online or offline using any desired correlation algorithm.

An image of a patterned GaAs chip (Fig. 1) acquired with a primary ion beam of I^{4+} ions at an energy of 15 MeV is shown in Fig. 2. To obtain the image the combined absolute intensity of three characteristic mass peaks around m/q = 43, 72 and 147 u/e was plotted.

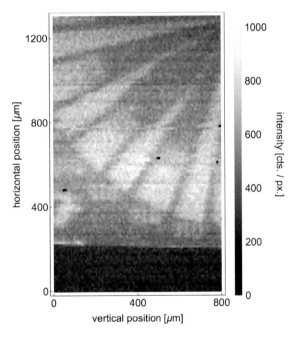

Fig. 2: *Combined secondary ion image of mass peaks around m/q = 43, 72 and 147 u/e obtained from the GaAs chip shown in Fig. 1.*

The used capillary has an outlet diameter of 5.4 µm at a working distance of ca. 30 mm from the sample surface. The pixels have a size of 10 µm^2 with a dwell time of 1 s. The resulting lateral resolution is approximately 15 µm (FWHM).

[1] M. Schulte-Borchers, Diss. ETH No. 23600 (2016)

[2] M. Schulte-Borchers et al., Nucl. Instr. and Meth. B 380 (2016) 94

MODIFIED CONFOCAL DC MAGNETRON SPUTTERING

The influence of ion bombardment on thin film growth

M. Trant[1], M.Fischer[1], K. Thorwarth[1], H.J. Hug[1], J. Patscheider[1], M. Döbeli

In this project the synthesis of ternary and quarternary thin films based on aluminium nitride is investigated with special attention to the optical properties and hardness of these materials. Aluminium nitride thin films exposed to varying ion flux densities during the growth process were deposited by reactive direct current (DC) magnetron sputtering at EMPA in Dübendorf.

Fig. 1: *Top view of the open deposition chamber and the electromagnetic coil installed around the substrate holder.*

In order to separate the influence of ion bombardment from the influence of film composition, detailed elemental analysis from ERDA and RBS measurements has been used. The plasma conditions at the substrate in the confocal DC magnetron sputtering setup are controlled by an electromagnetic coil (see Fig. 1 and 2). The additional magnetic field allows for a variation of low energy (<30 eV) ion flux density by more than one order of magnitude by altering the plasma confinement (Fig. 3). With this modified deposition setup the residual stress and the growth morphology of the thin film can be controlled. In addition it has been observed that the oxygen content in the Al-O-N films can be tuned by the ion flux density during deposition. Ion beam analysis provides the

elemental composition of the bulk with very high precision and a very low detection limit.

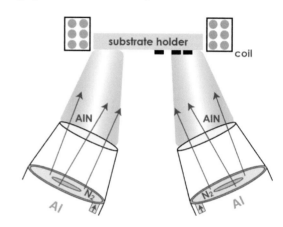

Fig. 2: *Schematics of the modified reactive DC magnetron sputtering setup. Black rectangles depict the substrates that are positioned at locations with different plasma conditions.*

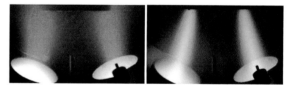

Fig. 3: *View of the sputter deposition chamber, with the substrate holder at the top. The effect of an additional magnetic field (right side) on the plasma confinement is clearly visible.*

For binary AlN films this proved especially valuable for the oxygen concentration resulting from residual gas inside the deposition chamber as measured by combining 2 MeV ^{4}He RBS and 13 MeV ^{127}I Heavy Ion ERDA. Furthermore the incorporation of argon due to increased ion bombardment could be excluded.

---

[1] *Nanoscale Materials Science, EMPA Dübendorf*

POROUS METAL FILMS

Chemical dealloying of a dense CuAl matrix observed by RBS

L. Dumée[1], B. Lin[1], H. Ma[2], R. Spolenak[2], M. Döbeli

Porous metal frameworks are promising materials with potential applications in a number of areas e.g. as catalytic reactors, membranes, electrodes, actuators, heat exchangers, or biocompatible prostheses [1]. Due to surface vacancies formed on the nano-textured metal surface they have specific advantages over similar bulk metals (Fig. 1).

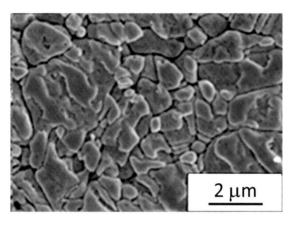

Fig. 1: *Micrograph of a nanoporous CuZn network dealloyed in 1 M NaOH solution.*

In this study, nanoporous materials were produced by chemical dealloying of metal films through either a recirculated or non-recirculated system (Fig. 2) dissolving Zn or Al.

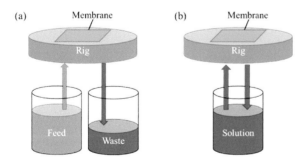

Fig. 2: *Schematic of the dealloying process either in non-recirculated (a) or recirculated (b) setup.*

The dealloying process of CuAl was observed by RBS which not only yields the total Al/Cu ratio but also reveals the concentration depth profile of Cu and Al across the film. The advance of the dealloying front from the surface towards the film interface as a function of time can be conveniently followed (Fig. 3).

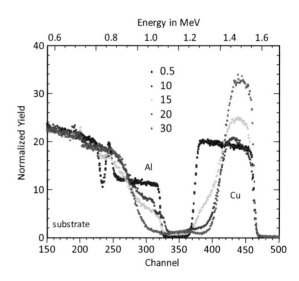

Fig. 3: *RBS spectra of initially 400 nm thick $Cu_{30}Al_{70}$ films. Dealloying time is given in minutes.*

This helps to understand the pore formation kinetics as well as the degree of atomic rearrangement upon selective etching of the alloy. The RBS results provide insights on the bulk composition of the material which could not be otherwise readily obtained by XPS analysis.

[1] Bao Lin et al., Nanomaterials 4 (2014) 856

[1] *Institute for Frontier Materials, Deakin University, Australia*
[2] *Department of Materials, ETH Zürich*

MECHANICAL PROPERTIES OF THIN FILM STACKS

Observation of intermetallic formation in W/Al multilayers by RBS

K. Stalder[1], S. Danzi[1], R. Spolenak[1], M. Döbeli

Metallic multilayers are well suited model structures to study and systematically analyze the role of interfaces in plastic deformation [1]. Therefore, the resulting lattice arrangement produced when two materials with significantly different crystal structures come together is of special interest, as it can significantly influence the mechanical system behavior. Such architectures are obtained in most cases through vapor phase processes, i.e. magnetron sputtering, as those techniques allow for a precise control over the single layer thickness. Tungsten/aluminum film stacks are promising candidates as the large shear modulus mismatch between the constituent metals might lead to increased hardness.

Multilayers of W and Al were deposited by PVD onto silicon substrates (Fig. 1) with single layer thicknesses between 4 and 40 nm. The samples were then annealed at temperatures between 400 and 600°C.

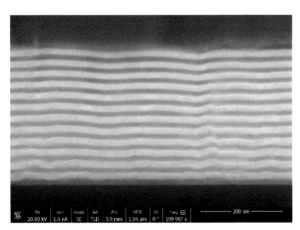

Fig. 1: *SEM cross section micrograph: W/Al as deposited sample.*

Composition depth profiles before and after annealing of the multilayer systems were measured by RBS (Fig. 2). Sublayers could be resolved down to a thickness of about 12 nm. RBS offers a fast means for tracking of interdiffusion and intermetallic formation at interfaces (Fig. 2, bottom). The observed intermixing can now be compared to the change in mechanical properties of the coatings.

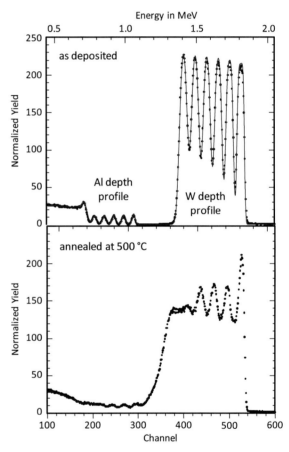

Fig. 2: *Multilayer system on Si with six periods of W (40 nm)/Al (45 nm). Top: as deposited, bottom: annealed at 500 °C.*

[1] A. Misra et al., Scripta Materialia 39 (1998) 555

[1] *Department of Materials, ETHZ*

HIGH BORON-CONTENT GRAPHITIC CARBON

Measuring boron to carbon ratios by MeV ion beam techniques

N. Stadie[1], E. Billeter[1], M. Kovalenko[1], M. Döbeli

The physical and electrochemical properties of graphite can be influenced by the substitution of carbon atoms within the graphitic lattice by neighboring elements like boron or nitrogen [1]. In this project the replacement of carbon by boron is achieved by a simple, single-step co-pyrolysis of miscible liquid precursors at 800 °C. The product is a bulk material in the form of dark lustrous flakes with a disordered graphitic structure (Fig. 1). The boron to carbon ratio has been determined by Heavy Ion ERDA with a 13 MeV [127]I primary ion beam (Fig. 2). Sample preparation for the ERDA measurements is somewhat challenging since a large flat sample area is needed and no carbon-containing glue or tape should be exposed to the beam.

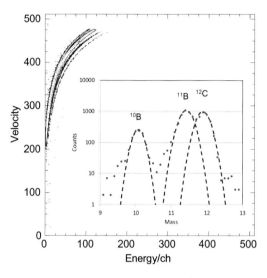

Fig. 2: *Biparametric 13 MeV [127]I Heavy Ion ERDA spectrum of the B_4C standard sample. The inset shows the extracted mass spectrum.*

Fig. 1: *Directly-synthesized B-doped graphitic carbon: (a) liquid precursor mixture, (b) post-pyrolysis product, and (c) harvested flakes.*

Therefore, low energy RBS at 1.4 MeV has been measured as well (Fig. 3). The low He energy is needed to avoid nuclear resonances and reactions with the light boron isotopes. The backscatter signal from ^{10}B and ^{11}B is at 300 keV only. Nevertheless, with a detector resolution of 13 keV all masses between 10 and 16 u could be separated and the B/C ratios could be reliably determined as long as there are no heavy contaminants in the material. Further synthetic effort to lower the B impurity level are underway to facilitate future He RBS.

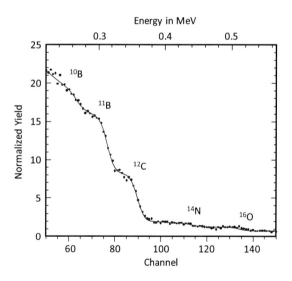

Fig. 3: *1.4 MeV RBS spectrum of the B_4C standard sample.*

[1]   C. E. Lowell, J Am. Ceram. Soc. 50 (1967) 142

[1] *Laboratory of Inorganic Chemistry, ETH Zurich*

TOTAL IBA

Superalloy composition analysis by combining RBS, ERDA, and PIXE

J. Ast [1], M. Gindrat[2], A. Domann[3], X. Maeder[1], A. Neels[4], J. Ramm[5], K. von Allmen[4], M. Döbeli

"Total IBA" is an expression coined almost 20 years ago describing the combination of several MeV Ion Beam Analysis techniques to determine the complete elemental content of a material.

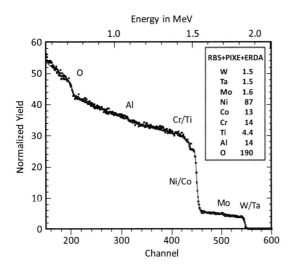

Fig. 1: *2 MeV He RBS spectrum of an oxidized superalloy sample. Final composition is given in at% of the Ni-Co matrix.*

Here we used heavy ion ERDA and simultaneous RBS and PIXE measurements to analyze a nickel superalloy oxide containing nine elements.

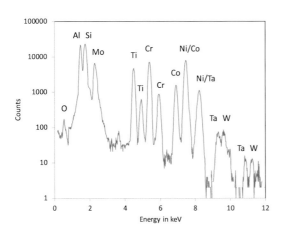

Fig. 2: *2 MeV He PIXE spectrum simultaneously taken with RBS data shown in Fig. 1.*

The ratio of light elements (O, Al) and heavy elements (Mo, Ta, W) to the main Ni-Co matrix was determined by RBS. Elements with similar isotopic masses not distinguishable by RBS (Ta-W and Ni-Co) were then discriminated by PIXE (Fig. 2). Since these pairs of elements have similar atomic numbers the systematic errors induced by X-ray production and absorption coefficients are greatly reduced.

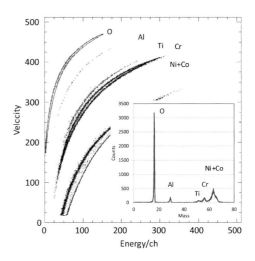

Fig. 3: *Two-dimensional ERDA data and mass spectrum of the same sample.*

Finally, the Cr/Ti ratio and the Al content were determined by 13 MeV I heavy ion ERDA (Fig. 3). These results (Fig. 1) complement intensive structural analysis of the same material by a variety of other techniques (XRD, SEM, EDX, TEM, FIB, etc.).

[1]   C. Jeynes et. al., Nucl. Instr. Methods Volume B271 (2012) 107

[1] *Empa Thun*
[2] *Oerlikon Metco AG, Wohlen*
[3] *Empa St. Gallen*
[4] *EMPA Dübendorf*
[5] *Oerlikon Surface Solutions AG, Balzers*

SELECTIVE GRAIN GROWTH BY ION-IRRADIATION

Converting polycrystalline tungsten thin films into single crystals

H. Ma[1], R. Spolenak[1], M. Döbeli

Material properties are closely related to their microstructure, e.g. grain size and texture. Conventional thermomechanical treatments are commonly applied to achieve specific features. The viability of this route is largely limited in thin films of refractory metals, due to restricted deformability and high melting temperature. We demonstrate an alternative method for microstructure engineering in metallic thin films by ion irradiation. 500 nm thick W films with (110) fiber texture, sputtered on Si wafers, are bombarded by 4.5 MeV Au ions at liquid nitrogen temperature with an angle of 35° to the sample surface normal, corresponding to the <111> channeling direction (Fig. 1). The damage and Au concentration profile calculated by SRIM are presented in Fig. 2.

inhomogeneous distribution of ion-induced defects between channeling and non-channeling oriented grains.

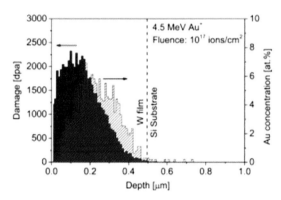

Fig. 2: *SRIM calculation of damage and Au concentration profile in W film on Si substrate with a fluence of 10^{17} cm^{-2} 4.5 MeV Au ions.*

Because of the low defect generation rate (i.e. low volume free energy) in channeling oriented grains, they grow at the expense of the remaining ones.

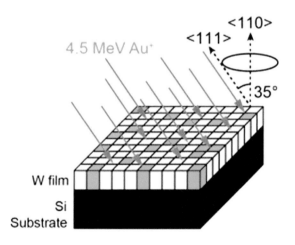

Fig. 1: *Illustration of irradiation configuration. Grains with one of their [111] directions parallel to the ion beam are highlighted in green.*

The microstructure evolution of W films with different texture sharpness upon ion-irradiation is characterized by EBSD (Electron Backscatter Diffraction), as shown in Fig. 3. Significant grain growth and development of in-plane texture is obtained by ion irradiation. The selectivity of grain growth is attributed to the

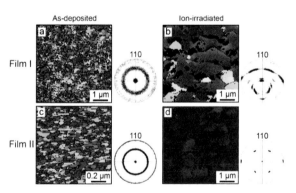

Fig. 3: *EBSD in-plane orientation maps and corresponding (110) pole figures of two W films with different texture sharpness before (a and c) and after (b and d) ion irradiation at 77K.*

---

[1] *Department of Materials, ETH Zurich*

PRODUCTION CROSS-SECTION OF Ca-42(d,n)Sc-43

A short-lived radioisotope for application in PET

N.P. van der Meulen[1], C. Vermeulen[1], A. Türler[1], A.M. Müller, M. Döbeli, H.-A. Synal

Positron Emission Tomography (PET) is one of the most widely used techniques to image metabolic processes in the body and is a standard diagnostic tool in oncology. ^{18}F is most frequently applied as the PET radiotracer of choice. ^{18}F is suitable to label small organic molecules, but has some disadvantages in labeling peptides or proteins. Radio-metals are more viable for these kinds of molecules. In recent years, ^{68}Ga, having a half-life of 1.13 h has risen in prominence. ^{43}Sc is a very promising radio-metal in that it is a better analogue for the therapeutic radionuclide ^{117}Lu than ^{68}Ga and it has a more favorable half-life at 3.89 h. ^{43}Sc can be produced by proton or deuteron irradiation of enriched calcium isotopes. In this project we measured the reaction cross-section of the reaction $^{42}Ca(d,n)^{43}Sc$ with deuteron beams from 6 to 8.5 MeV (Fig. 1).

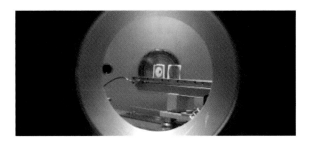

Fig. 1: *Beam position and CaCO₃ target.*

The signature of ^{43}Sc could well be identified in the γ-spectrum of the activated material (Fig. 2). The preliminary results for the excitation curve of the nuclear reaction $^{42}Ca(d,n)^{43}Sc$ is presented in Fig. 3. Compared to the results of De Waal et al [1] and the TENDL [2] library. Our results seem to be in the right magnitude but shifted in energy.

Further refinement of the bombardment setup and better target deposition methods are necessary to eliminate large uncertainties and pin down the cross-section in the available deuteron energy range.

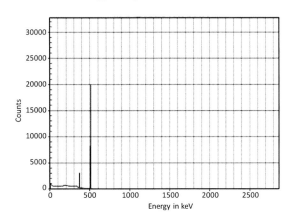

Fig. 2: *Gamma spectrum of ^{43}Sc.*

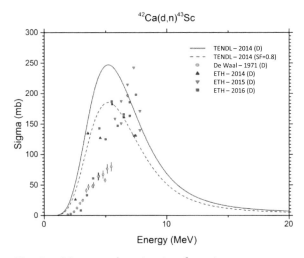

Fig. 3: *Measured excitation function.*

[1] N.P. van der Meulen et al., Nucl. Med. Biol. 42 (2015) 74

[2] TENDL-2015: TALYS-based evaluated nuclear data library, A.J. Koning, et.al, https://tendl.web.psi.ch/tendl_2015/tendl 2015.htm

---

[1] *Radiochemistry, Paul Scherrer Institut, Villigen*

EDUCATION

School projects in 2016

Reconstructing Palaeoglaciers with GIS

Development of new equipment for the LIP

SCHOOL PROJECTS IN 2016

Leaves, tooth and Ötzi—a wide range of ^{14}C ages

I. Hajdas, M. Maurer, M.B. Röttig, H.-A. Synal, L. Wacker

During the last 9 years at least 2 projects involving high school students were completed each year. Samples from the archived tissue of Ötzi, the Ice Man (Fig. 1) were analyzed again as a part of the ETH project: *Studienwoche* as well as for a school project of Kantonsschule Olten. Frozen tissue of Ötzi was treated with acid-base-acid solutions to remove contamination with carbonates and humic acids [2]. The results confirmed the age of 4550 ± 27 BP obtained by ETH laboratory 25 years ago [1].

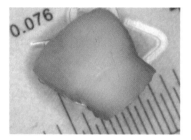

Fig. 2: *Milk tooth used for ^{14}C analysis.*

In addition leaves of shrubs growing at Hönggerberg were analyzed to measure present day atmospheric ^{14}C content. The trend of the atmosphere becoming 'older' due to addition of 'old' CO_2 from combustion of fossil (^{14}C free) carbon is visible in figure 3.

Fig. 1: *Fragments of an archived Ötzi tissue stored frozen in laboratory archives since September 1991.*

One of the students who participated in the ETH project 'sacrificed' her own milk tooth (Fig. 2). This material required a little different preparation of the organic matter than the frozen tissue. Bones and teeth contain organic matter in form of collagen, which must be extracted from the sample material. The results of AMS analysis on the collagen show that the tooth was formed between 1997 and 2001. The student was born in 1999 (Fig. 3).

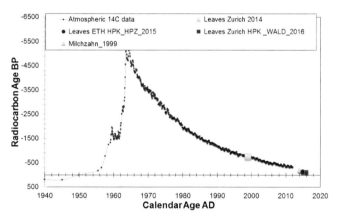

Fig. 3: *Dating of a 'baby' tooth to years 1997-2001 (Blue Square) corresponds to the year of birth 1999. Concentration of ^{14}C in fresh leaves (green, red and brown) shows effect of 'old' carbon added to the atmosphere.*

In summary, the school projects provided opportunity for school students to work on real samples and produce valuable data.

[1] G. Bonani et al., Radiocarbon 36 (1994) 247
[2] I. Hajdas, Quaternary Science Journal - Eiszeitalter und Gegenwart 57 (2008) 2

RECONSTRUCTING PALAEOGLACIERS WITH GIS

A workshop on new tools to calculate palaeoglacier surfaces and ELAs

S. Ivy-Ochs, R. Pellitero[1], K. Hippe

We organized a small workshop (Fig. 1) to familiarize BSc, MSc and PhD students with the new toolboxes developed for the use in the geographic information system (GIS) software ArcGIS. To lead the workshop, we invited Dr. R. Pellitero from the University of Aberdeen, the developer of the tools. The toolboxes are used i) to reconstruct palaeoglacier surfaces and ii) to calculate of the equilibrium-line altitude (ELA) of the reconstructed glacier.

Fig. 1: *Participants of the workshop.*

The ELA of a palaeoglacier is a key parameter for understanding past climates. Its knowledge allows inferences to be made about temperature and precipitation at the time of the glacier advance. Within the specific research of the participants, palaeoglacier extents are constrained by mapping of moraines, while the timing is constrained with cosmogenic nuclide exposure dating. For their study sites the students reconstructed ice surfaces using the GlaRe toolbox [1]. This toolbox provides individual tools to generate the ice thickness from the bed topography along a palaeoglacier flowline applying the standard flow law for ice. It generates the 3D surface of a palaeoglacier using multiple interpolation methods.

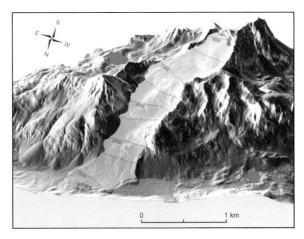

Fig. 2: *Reconstruction of a palaeoglacier at Julier Pass, Switzerland, using the GlaRE toolbox for ArcGIS [3].*

Next the ELA of the reconstructed glacier was calculated with the second toolbox [2]. This toolbox allows the automated calculation of glacier equilibrium-line altitudes using the Accumulation Area Ratio, Area-Altitude Balance Ratio. Using the reconstructed surface of the palaeoglacier as input, the GIS-based ELA calculation can be done for single glaciers as well as for large dataset with several glaciers.

[1] R. Pellitero et al., Computers and Geosci. 82 (2015) 55

[2] R. Pellitero et al., Computers and Geosci. 94 (2016) 77

[3] S. Ivy-Ochs, Cuadernos de investigación geográfica 41 (2015) 295

[1] *Geography and Environment, University of Aberdeen, U.K.*

DEVELOPMENT OF NEW EQUIPMENT FOR THE LIP

Apprentices design and assemble new tools

L. Wacker, K. Seidler, A. Zvyagin

The LIP educates every year apprentices in physics. Last year, two students in the third year developed a series of new devices for the LIP that are now in use.

A first project to develop a high-precision magnetic field measurement device (layout see Fig. 1) based on a cheap, commercially available hall sensor (HE244, Asensor AB, Sweden) was accomplished by Alexander Zvyagin and Karl Seidler. The device can read out and display two hall probes and transfer the measured values over a serial port to a PC. The readout in the range of 0.2 to 1 T is stable to about 0.1‰ over a day and shows a low temperature dependency. The device is now operational and its long-term stability is being tested on the MICADAS.

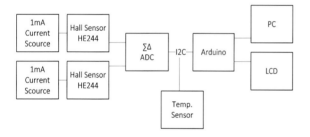

Fig. 1: *Schematic layout of the developed hall probe setup.*

In a second project Karl developed a tube cracker for 8 mm quartz tubes. The device allows to transfer CO_2 from tube combusted samples to the graphitization line AGE (*Fig. 2*). The samples are cracked in a cracker tube from where the CO_2 is transferred over a water trap to the zeolite trap of the AGE graphitization line.

Fig. 2: *Tube cracker system for the automated graphitization line AGE.*

Finally, Karl also designed a machine for the separation of individual tree rings from cores drilled from trunks of trees (*Fig. 3*). Individual tree rings are cut with a razor blade mounted to a handle bar with liners. The device allows to adapt the razor blade to the orientation and shape of the tree-rings in a drill core.

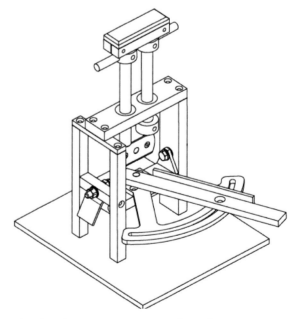

Fig. 3: *Ring separation machine for cutting tree rings from a drill core.*

PUBLICATIONS

N. Akçar, V. Alfimov, S. Ivy Ochs, V. Yavuz, I. Ö. Yilmaz, D. Altiner and C. Schlüchter
Ausgrabungen und Forschungen in der Westlichen Oberstadt von Hattusa I
DEUTSCHES ARCHÄOLOGISCHES INSTITUT (2016) 69-87

R. Bao, C. McIntyre, M. Zhao, C. Zhu, S.-J. Kao and T. I. Eglinton
Widespread dispersal and aging of organic carbon in shallow marginal seas
Geology **44** (2016) 791-794

M. G. Bichler, M. Reindl, J. M. Reitner, R. Drescher-Schneider, C. Wirsig, M. Christl, I. Hajda and
S. Ivy Ochs
Landslide deposits as stratigraphical markers for a sequence-based glacial stratigraphy: a case study of a Younger Dryas system in the Eastern Alps
BOREAS - An international journal of Quaternary research (2016) 1-15

J. Braakhekke, F. Kober and S. Ivy Ochs
A GIS-based compilation of the erratic boulders of the Last Glacial Maximum in the Rhine Glacier system
National
Cooperative for the Disposal of Radioactive Waste Management and Environmental Restoration (2016)

U. Büntgen, V. S. Myglan, F. C. Ljungqvist, M. McCormick, N. Di Cosmo, M. Sigl, J. Jungclaus,
S. Wagner, P. J. Krusic, J. Esper, J. O. Kaplan, M. A. C. De Vaan, J. Luterbacher, L. Wacker, W. Tegel and A. V. Kirdyanov
Cooling and societal change during the Late Antique Little Ice Age from 536 to around 660 AD
Nature Geoscience **9** (2016) 231-237

N. Casacuberta, P. Masque, G. Henderson, M. R. van-der-Loeff, D. Bauch, C. Vockenhuber, A. Daraoui, C. Walther, H. A. Synal and M. Christl
First ^{236}U data from the Arctic Ocean and use of $^{236}U/^{238}U$ and $^{129}I/^{236}U$ as a new dual tracer Earth and Planetary Science Letters **440** (2016) 127-134

J. Chen, M. Döbeli, D. Stender, M. Lee, K. Conder, C. Schneider, A. Wokaun and T. Lippert
Tracing the origin of oxygen for $La_{0.6}Sr_{0.4}MnO_3$ thin film growth by pulsed laser deposition
Journal of Physics D: Applied Physics **49** (2016) 1-10

R. Cusnir, M. Jaccard, C. Bailat, M. Christl, P. Steinmann, M. Haldimann, F. Bochud and P. Froidevaux
Probing the kinetic parameters of Plutonium–Naturally occurring organic matter interactions in freshwaters using the diffusive gradients in thin films technique
Environmental Science & Technology **50** (2016) 5103-5110

X. Dai, M. Christl, S. Kramer-Tremblay and H.-A. Synal
Determination of Atto-to Femtogram Levels of Americium and Curium Isotopes in Large-Volume Urine Samples by Compact Accelerator Mass Spectrometry
Analytical Chemistry **88** (2016) 2832-2837

A. Daraoui, B. Riebe, C. Walther, H. Wershofen, C. Schlosser, C. Vockenhuber and H.-A. Synal
Concentrations of iodine isotopes (^{129}I and ^{127}I) and their isotopic ratios in aerosol samples from Northern Germany
Journal of Environmental Radioactivity **154** (2016) 101-108

J. L. Dixon, F. von Blanckenburg, K. Stüwe and M. Christl
Glaciation's topographic control on Holocene erosion at the eastern edge of the Alps
Earth Surface Dynamics **4** (2016) 895-909

M. Dühnforth, A. L. Densmore, S. Ivy-Ochs, P. Allen and P. W. Kubik
Early to Late Pleistocene history of debris-flow fan evolution in western Death Valley using cosmogenic ^{10}Be and ^{26}Al
Geomorphology **281** (2016) 53-65

U. Dusek, R. Hitzenberger, A. Kasper-Giebl, M. Kistler, H. A. J. Meijer, S. Szidat, L. Wacker, R. Holzinger and T. Röckmann
Sources and formation mechanisms of carbonaceous aerosol at a regional background site in the Netherlands: Insights from a year-long radiocarbon study
Atmospheric Chemistry and Physics (2016) 1-40

B. Frey, T. Rime, M. Phillips, B. Stierli, I. Hajdas, F. Widmer and M. Hartmann
Microbial diversity in European alpine permafrost and active layers
FEMS Microbiology Ecology **92** (2016) 1-17

M. Gatta, G. Sinopoli, M. Giardini, B. Giaccio, I. Hajdas, L. Pandolfi, G. Bailey, P. Spikins, M. F. Rolfo and L. Sadori
Pollen from Late Pleistocene hyena (Crocuta crocuta spelaea) coprolites: An interdisciplinary approach from two Italian sites
Review of Palaeobotany and Palynology **233** (2016) 56-66

M. Gierga, I. Hajdas, U. J. van Raden, A. Gilli, L. Wacker, M. Sturm, S. M. Bernasconi and R. H. Smittenberg
Long-stored soil carbon released by prehistoric land use: Evidence from compound-specific radiocarbon analysis on Soppensee lake sediments
Quaternary Science Reviews **144** (2016) 123-131

L. M. Grämiger, J. R. Moore, C. Vockenhuber, J. Aaron, I. Hajdas and S. Ivy-Ochs
Two early Holocene rock avalanches in the Bernese Alps (Rinderhorn, Switzerland)
Geomorphology **268** (2016) 207-221

M. E. Habicht, R. Bianucci, S. A. Buckley, J. Fletcher, A. S. Bouwman, L. M. Öhrström, R. Seiler, F. M. Galassi, I. Hajdas, E. Vassilika, T. Böni, M. Henneberg and F. J. Rühli
Queen Nefertari, the Royal Spouse of Pharaoh Ramses II: A Multidisciplinary Investigation of the Mummified Remains Found in Her Tomb (QV66)
Plos One **11** (2016) 1-20

K. Häusler, M. Moros, L. Wacker, L. Hammerschmidt, O. Dellwig, T. Leipe, A. Kotilainen and H. W. Arz
Mid-to late Holocene environmental separation of the northern and central Baltic Sea basins in response to differential land uplift
Boreas (2016) 1-18

L. Hendriks, I. Hajdas, C. McIntyre, M. Küffner, N. C. Scherrer and E. S. B. Ferreira
Microscale radiocarbon dating of paintings
Applied Physics A **122** (2016) 1-10

A. Hogg, J. Southon, C. Turney, J. Palmer, C. B. Ramsey, P. Fenwick, G. Boswijk, U. Büntgen,
M. Friedrich, G. Helle, K. Hughen, R. Jones, B. Kromer, A. Noronha, F. Reinig, L. Reynard, R. Staff and L. Wacker
Decadally Resolved Lateglacial Radiocarbon Evidence from New Zealand Kauri
Radiocarbon (2016) 1-25

A. Hogg, J. Southon, C. Turney, J. Palmer, C. B. Ramsey, P. Fenwick, G. Boswijk, M. Friedrich, G. Helle, K. Hughen, R. Jones, B. Kromer, A. Noronha, L. Reynard, R. Staff and L. Wacker
Punctuated Shutdown of Atlantic Meridional Overturning Circulation during Greenland Stadial 1
Scientific Reports **6** (2016) 1-10

E. Huysecom, I. Hajdas, M.-A. Renold, H.-A. Synal and A. Mayor
The "Enhancement" of Cultural Heritage by AMS Dating: Ethical Questions and Practical Proposals
Radiocarbon (2016) 1-5

P. Irkhin, I. Biaggio, T. Zimmerling, M. Döbeli and B. Batlogg
Defect density dependent photoluminescence yield and triplet diffusion length in rubrene
Applied Physics Letters **108** (2016)

S. Kusch, J. Rethemeyer, E. C. Hopmans, L. Wacker and G. Mollenhauer
Factors influencing ^{14}C concentrations of algal and archaeal lipids and their associated sea surface temperature proxies in the Black Sea
Geochimica Et Cosmochimica Acta **188** (2016) 35-57

B. Lebrun, T. Chantal, C. Benoît, R. Michel, L. Laurent, L. Alice, H. Irka, C. Abdoulaye, M. Norbert and H. Éric
Establishing a West African chrono-cultural framework: First luminescence dating of sedimentary formations from the Falémé Valley, Eastern Senegal
Journal of Archaeological Science: Reports **7** (2016) 379-388

F. A. Lechleitner, J. U. Baldini, S. F. Breitenbach, J. Fohlmeister, C. McIntyre, B. Goswami,
R. A. Jamieson, T. S. van der Voort, K. Prufer, N. Marwan, B. J. Culleton, D. J. Kennett, Y. Asmerom,
V. Polyak and T. I. Eglinton
Hydrological and climatological controls on radiocarbon concentrations in a tropical stalagmite
Geochimica Et Cosmochimica Acta **194** (2016) 233-252

F. A. Lechleitner, J. Fohlmeister, C. McIntyre, L. M. Baldini, R. A. Jamieson, H. Hercman,
M. Gąsiorowski, J. Pawlak, K. Stefaniak, P. Socha, T. I. Eglinton and J. U. L. Baldini
A novel approach for construction of radiocarbon-based chronologies for speleothems Quaternary Geochronology **35** (2016) 54-66

J. Lippold, M. Gutjahr, P. Blaser, E. Christner, M. L. de Carvalho Ferreira, S. Mulitza, M. Christl,
F. Wombacher, E. Böhm and B. Antz
Deep water provenance and dynamics of the (de) glacial Atlantic meridional overturning circulation
Earth and Planetary Science Letters **445** (2016) 68-78

X. Maeder, A. Neels, M. Döbeli, A. Dommann, H. Rudigier, B. Widrig and J. Ramm
Comparison of in-situ oxide formation and post-deposition high temperature oxidation of Ni-aluminides synthesized by cathodic arc evaporation
Surface and Coatings Technology **309** (2016) 516-522

C. P. McIntyre, F. Lechleitner, S. Q. Lang, N. Haghiour, S. Fahrni, L. Wacker and H.-A. Synal
[14]C Contamination Testing in Natural Abundance Laboratories: A New Preparation Method Using Wet Chemical Oxidation and Some Experiences
Radiocarbon **58** (2016) 935-941

C. P. McIntyre, L. Wacker, N. Haghipour, T. M. Blattmann, S. Fahrni, M. Usman, T. I. Eglinton and H.-A. Synal
ONLINE ^{13}C AND ^{14}C GAS MEASUREMENTS BY EA-IRMS–AMS AT ETH ZÜRICH
Radiocarbon (2016) 1-11

A. P. Moran, S. Ivy-Ochs, M. Schuh, M. Christl and H. Kerschner
Evidence of central Alpine glacier advances during the Younger Dryas–early Holocene transition period
Boreas (2016) 1-10

A. P. Moran, S. I. Ochs, C. Vockenhuber and H. Kerschner
Rock glacier development in the Northern Calcareous Alps at the Pleistocene-Holocene boundary
Geomorphology **273** (2016) 178-188

A. E. Ojala, L. Arppe, T. P. Luoto, L. Wacker, E. Kurki, M. Zajączkowski, J. Pawłowska, M. Damrat and M. Oksman
Sedimentary environment, lithostratigraphy and dating of sediment sequences from Arctic lakes Revvatnet and Svartvatnet in Hornsund, Svalbard
Polish Polar Research **37** (2016) 23-48

A. Ojeda-GP, C. W. Schneider, M. Döbeli, T. Lippert and A. Wokaun
The importance of pressure and mass ratios when depositing multi-element oxide thin films by pulsed laser deposition
Applied Surface Science **389** (2016) 126-134

A. A. Osman, S. Bister, B. Riebe, A. Daraoui, C. Vockenhuber, L. Wacker and C. Walther
Radioecological investigation of ^{3}H, ^{14}C, and [129]I in natural waters from Fuhrberger Feld catchment, Northern Germany
Journal of Environmental Radioactivity **165** (2016) 243-252

R. Pellitero, B. R. Rea, M. Spagnolo, J. Bakke, S. Ivy-Ochs, C. R. Frew, P. Hughes, A. Ribolini, S. Lukas and H. Renssen
Glare, a GIS tool to reconstruct the 3D surface of palaeoglaciers
Computers & Geosciences **94** (2016) 77 - 85

K. Perner, A. E. Jennings, M. Moros, J. T. Andrews and L. Wacker
Interaction between warm Atlantic-sourced waters and the East Greenland Current in northern Denmark Strait (68° N) during the last 10 600 cal a BP
Journal of Quaternary Science **31** (2016) 472-483

M. Petrillo, P. Cherubini, G. Fravolini, M. Marchetti, J. Ascher-Jenull, M. Schärer, H.-A. Synal,
D. Bertoldi, F. Camin and R. Larcher
Time since death and decay rate constants of Norway spruce and European larch deadwood in subalpine forests determined using dendrochronology and radiocarbon dating
Biogeosciences **13** (2016) 1537-1552

M. Pichler, D. Pergolesi, S. Landsmann, V. Chawla, J. Michler, M. Döbeli, A. Wokaun and T. Lippert
TiN-buffered substrates for photoelectrochemical measurements of oxynitride thin films Applied
Surface Science **369** (2016) 67-75

A. E. Putnam, D. E. Putnam, L. Andreu-Hayles, E. R. Cook, J. G. Palmer, E. H. Clark, C. Wang, F. Chen, G. H. Denton, D. P. Boyle, S. D. Bassett, S. D. Birkel, J. Martin-Fernandez, I. Hajdas, J. Southon,
C. B. Garner, H. Cheng and W. S. Broecker
Little Ice Age wetting of interior Asian deserts and the rise of the Mongol Empire
Quaternary Science Reviews **131** (2016) 33-50

J. M. Reitner, S. Ivy Ochs, R. Drescher-Schneider, I. Hajda and M. Linner
Reconsidering the current stratigraphy of the Alpine Lateglacial: Implications of the sedimentary and morphological record of the Lienz area (Tyrol/Austria)
Quaternary Science Journal **65** (2016) 113 - 144

T. Ronge, R. Tiedemann, F. Lamy, P. Köhler, B. Alloway, R. De Pol-Holz, K. Pahnke, J. Southon and
L. Wacker
Radiocarbon constraints on the extent and evolution of the South Pacific glacial carbon pool
NATURE COMMUNICATIONS **7** (2016)

M. Salehpour, K. Håkansson, G. Possnert, L. Wacker and H.-A. Synal
Performance report for the low energy compact radiocarbon accelerator mass spectrometer at Uppsala University
Nuclear Instruments and Methods in Physics Research Section B: Beam Interactions with Materials and Atoms **371** (2016) 360-364

M. Schaller, T. Ehlers, T. Stor, J. Torrent, L. Lobato, M. Christl and C. Vockenhuber
Spatial and temporal variations in denudation rates derived from cosmogenic nuclides in four European fluvial terrace sequences
Geomorphology **274** (2016) 180-192

M. Schaller, T. A. Ehlers, T. Stor, J. Torrent, L. Lobato, M. Christl and C. Vockenhuber
Timing of European fluvial terrace formation and incision rates constrained by cosmogenic nuclide dating
Earth and Planetary Science Letters **451** (2016) 221-231

J. Schoonejans, V. Vanacker, S. Opfergelt, Y. Ameijeiras-Mariño and M. Christl
Kinetically limited weathering at low denudation rates in semi-arid climatic conditions
Journal of Geophysical Research: Earth Surface (2016) 1-15

M. Schulte-Borchers, M. Döbeli, A. M. Müller, M. George and H.-A. Synal
Time-of-flight MeV-SIMS with beam induced secondary electron trigger
Nuclear Instruments and Methods in Physics Research Section B: Beam Interactions with Materials and Atoms **380** (2016) 94-98

P. Schürch, A. L. Densmore, S. Ivy-Ochs, N. J. Rosser, F. Kober, F. Schlunegger, B. McArdell and V. Alfimov
Quantitative reconstruction of late Holocene surface evolution on an alpine debris-flow fan
Geomorphology **275** (2016) 46-57

G. Scognamiglio, E. Chamizo, J. López-Gutiérrez, A. Müller, S. Padilla, F. Santos, M. López-Lora, C. Vivo-Vilches and M. García-León
Recent developments of the 1MV AMS facility at the Centro Nacional de Aceleradores
Nuclear Instruments and Methods in Physics Research Section B: Beam Interactions with Materials and Atoms **375** (2016) 17-25

D. Scopece, M. Döbeli, D. Passerone, X. Maeder, A. Neels, B. Widrig, A. Dommann, U. Müller and J. Ramm
Silicon etch with chromium ions generated by a filtered or non-filtered cathodic arc discharge
Science and Technology of Advanced Materials **17** (2016) 20-28

K. Sen, E. Perret, A. Alberca, M. A. Uribe-Laverde, I. Marozau, M. Yazdi-Rizi, B. P. P. Mallett, P. Marsik, C. Piamonteze, Y. Khaydukov, M. Dobeli, T. Keller, N. Biskup, M. Varela, J. Vasatko, D. Munzar and C. Bernhard
X-ray absorption study of the ferromagnetic Cu moment at the $YBa_2Cu_3O_7/La_{2/3}Ca_{1/3}MnO_3$ interface and variation of its exchange interaction with the Mn moment
Physical Review B **93** (2016)

K. Shimamoto, M. Döbeli, T. Lippert and C. Schneider
Cation ratio and ferroelectric properties of $TbMnO_3$ epitaxial films grown by pulsed laser deposition
Journal of Applied Physics **119** (2016)

A. Sookdeo, L. Wacker, S. Fahrni, C. P. McIntyre, M. Friedrich, F. Reinig, D. Nievergelt, W. Tegel, B. Kromer and U. Büntgen
Speed Dating: A Rapid Way to Determine the Radiocarbon Age of Wood by EA-AMS
Radiocarbon (2016) 1-7

R. B. Sparkes, A. D. Selver, Ö. Gustafsson, I. P. Semiletov, N. Haghipour, L. Wacker, T. I. Eglinton, H. M. Talbot and B. E. van Dongen
Macromolecular composition of terrestrial and marine organic matter in sediments across the East Siberian Arctic Shelf
The Cryosphere **10** (2016) 2485-2500

V. Stoytschew, M. Schulte-Borchers, I. B. Mihalića and R. Perez
New type of capillary for use as ion beam collimator and air-vacuum interface
Nuclear Instruments and Methods in Physics Research Section B: Beam Interactions with Materials and Atoms **380** (2016) 99-102

S. Tao, T. I. Eglinton, D. B. Montluçon, C. McIntyre and M. Zhao
Diverse origins and pre-depositional histories of organic matter in contemporary Chinese marginal sea sediments
Geochimica Et Cosmochimica Acta **191** (2016) 70-88

T. S. v. d. Voort, F. Hagedorn, C. McIntyre, C. Zell, L. Walthert, P. Schleppi, X. Feng and T. I. Eglinton
Variability in ^{14}C contents of soil organic matter at the plot and regional scale across climatic and geologic gradients
Biogeosciences **13** (2016) 3427-3439

C. Welte, L. Wacker, B. Hattendorf, M. Christl, J. Fohlmeister, S. F. Breitenbach, L. F. Robinson, A. H. Andrews, A. Freiwald and J. R. Farmer
Laser Ablation–Accelerator Mass Spectrometry: An Approach for Rapid Radiocarbon Analyses of Carbonate Archives at High Spatial Resolution
Analytical Chemistry **88** (2016) 8570-8576

C. Wirsig, S. Ivy-Ochs, N. Akçar, M. Lupker, K. Hippe, L. Wacker, C. Vockenhuber and C. Schlüchter
Combined cosmogenic ^{10}Be, in situ ^{14}C and ^{36}Cl concentrations constrain Holocene history and erosion depth of Grueben glacier (CH)
Swiss Journal of Geosciences **109** (2016) 379-388

C. Wirsig, J. Zasadni, M. Christl, N. Akçar and S. Ivy-Ochs
Dating the onset of LGM ice surface lowering in the High Alps
Quaternary Science Reviews **143** (2016) 37-50

C. Wirsig, J. Zasadni, S. Ivy-Ochs, M. Christl, F. Kober and C. Schlüchter
A deglaciation model of the Oberhasli, Switzerland
Journal of Quaternary Science **31** (2016) 46-59

B. Zollinger, C. Alewell, C. Kneisel, D. Brandová, M. Petrillo, M. Plötze, M. Christl and M. Egli
Soil formation and weathering in a permafrost environment of the Swiss Alps: a multi-parameter and non-steady-state approach
Earth Surface Processes and Landforms (2016)

TALKS AND POSTERS

N. Akçar, S. Ivy-Ochs, V. Alfimov, A. Claude, R. Reber, M. Christl, C. Vockenhuber, F. Schlunegger, A. Dehnert, M. Rahn, C. Schlüchter
Isochron-burial dating of glacially-driven sediments: first results from the Alps
Austria, Vienna, 21.04.2016, EGU General Assembly 2016

N. Akçar, V. Yavuz, S. Ivy-Ochs, F. Nyffenegger, O. Fredin, M. Stolz, F. Schlunegger
February 2011 sensitive clay landslides at the Çöllolar coalfield, eastern Turkey
Austria, Vienna, 18.04.2016, EGU General Assembly 2016

S. Aksay, S. Ivy-Ochs, K. Hippe, L. Graemiger, C. Vockenhuber
The Geomorphological Evolution of a Landscape in a Tectonically Active Region: the Sennwald Landslide
Austria, Vienna, 21.04.2016, EGU General Assembly 2016

J. Anjar, N. Akçar, E. Larsen, A. Lyså, C. Vockenhuber
Dating the glacial activity on Jan Mayen using cosmogenic surface exposure dating with ^{36}Cl
Norway, Trondheim, 26.05.2016, 4th PAST Gateways International Conference

J. Ànspach, M. Lupker, N. Haghipour , T. I. Eglinton
Biomarker Signature of Greenland Sediments: from modern Rivers and Soils to MIS 5e and 11 Records
USA San francisco, 12.12.2016, AGU conference

R. Bao, N. Haghipour, L. Wacker, D. B. Montlucon, M. Zhao, A. Mcnichol, V. Galy , T. I.
EglintonRadiocarbon constraints on timescales of particulate organic matter transport over continental shelves
Japan, Yokohama, 26.06.2016, Goldschmidt 2016

J. Anjar, N. Akçar, E. Larsen, A. Lyså, C. Vockenhuber
Cosmogenic surface exposure dating with ^{36}Cl on Jan Mayen
Finland, Helsinki, 13.01.2016, 32nd Nordic Geological Winter Meeting

J. Braakhekke, S. Ivy-Ochs, I. Hajdas, G. Monegato, F. Gianotti, M. Christl
The Last Glacial Maximum around Lago d'Orta, Northern Italy; a multi method reconstruction
Austria, Vienna, 21.04.2016, EGU General Assembly 2016

J. Braakhekke, S. Ivy-Ochs, I. Hajdas, G. Monegato, F. Gianotti, M. Christl
The Last Glacial Maximum around Lago d'Orta, Northern Italy; a multi method reconstruction
Switzerland, Geneva, 19.11.2016, 14th Swiss Geoscience Meeting

N. Casacuberta, M. Christl, C. Vockenhuber, J. Vives-Batlle, P. Masqué, K. Buesseler
Assessment of the distribution of radionuclides (^{137}Cs, ^{134}Cs, ^{90}Sr, ^{129}I, ^{236}U and Pu-isotopes) in the coast off Japan derived from the Fukushima Dai-ichi nuclear accident
USA, San Diego, 03.11.2016, PICES - 2016

N. Casacuberta, P. Masqué, G. Henderson, M. Rutgers van der Loeff, D. Bauch, W. Walther, C. Vockenhuber, H.-A. Synal, M. Christl
Artificial radionuclides in the Arctic and North Atlantic Ocean
USA, New Orleans, 23.02.2016, Ocean Science Meeting

N. Casacuberta, K.O. Buesseler, J. Nishikawa, M. Christl, M. Aoyama, J. Vives-Batlle, V. Sanial, M.A. Charette, M. Castrillejo, P. Masqué
The impact of recent releases from the Fukushima nucelar accident on the marine environment
Spain, Sevilla, 08.11.2016, II International conference radioecological processes

N. Casacuberta, M. Castrillejo, M. Christl, C. Vockenhuber, X. Juan, H.-A. Synal, P. Masqué, K.O. Buesseler*Surface concentrations of ^{90}Sr, ^{129}I, and actinides measured in coastal waters off Japan 2-3 years after the Fukushima Dai-ichi nuclear accident*
Japan, Yokohama, 01.07.2016, Goldschmidt 2016

N. Casacuberta, P. Masqué, G. Henderson, M. Rutgers van der Loeff, D. Bauch, W. Walther, C. Vockenhuber, H.-A. Synal, M. Christl
^{236}U in the Arctic Ocean and implications of using $^{236}U/^{238}U$ and $^{129}I/^{236}U$ as a new dual tracer
Germany, Kiel, 26.01.2016, Siberian Shelf Workshop

N. Casacuberta
Artificial radionucies: an example of chemical tracers in oceanography
Switzerland, Zürich, 06.07.2016, AMS Seminar

M. Christl, N. Casacuberta, C. Vockenhuber, C. Elsässer, P. Bailly du Bois, J. Herrmann, H.-A. Synal
Reconstruction of The $^{129}I/^{236}U$ input function and its application for Transient tracer studies in the North Atlantic and Arctic Ocean
Hungary, Budapest, 12.04.2016, Conference on Radioanalytical and Nuclear Chemistry

A. Claude, N. Akçar, S. Ivy-Ochs, F. Schlunegger, P.W. Kubik, M. Christl, C. Vockenhuber, A. Dehnert, J. Kuhlemann, R. Meinert, C. Schlüchter
Landscape evolution of the northern Alpine Foreland: constructing a temporal framework for early to middle Pleistocene glaciations
Switzerland, Geneva, 19.11.2016, 14th Swiss Geoscience Meeting

A. Claude, N. Akçar, F. Schlunegger, S. Ivy-Ochs, P. Kubik, M. Christl, C. Vockenhuber, A. Dehnert, J. Kuhlemann, M. Rahn, C. Schlüchter
Landscape evolution and bedrock incision in the northern Alpine Foreland since the last 2 Ma
Austria, Vienna, 20.04.2016, EGU General Assembly 2016

A. Claude, N. Akçar, S. Ivy-Ochs, F. Schlunegger, P. Kubik, M. Christl, C. Vockenhuber, A. Dehnert, M. Rahn, C. Schlüchter
^{10}Be depth-profile dating of glaciofluvial sediments in the northern Alpine Foreland
Austria, Vienna, 21.04.2016, EGU General Assembly 2016

A. Cogez, F. Herman, E. Pelt, K. Norton, C. Darvill, M. Christl, G. Morvan, T. Reuschlé, F. Chabaux
U-Th and 10Be constraints on sediment recycling in proglacial settings, Lago Buenos Aires, Patagonia
Austria, Vienna, 20.04.2016, EGU General Assembly 2016

X. Dai, M. Christl, S. Kramer-Tremblay, H.-A. Synal
Ultra-sensitive Analysis of Actinides using compact accelerator Mass spectrometry
Hungary, Budapest, 13.04.2016, Conference on Radioanalytical and Nuclear Chemistry

M. Döbeli
Quantitative surface analysis by MeV ion beam techniques
Switzerland, Dübendorf, 29.06.2016, Surfaces and Thin Films - Analytics in Practice

M. Döbeli, K.-U. Miltenberger, A.M. Müller, M. George, H.-A. Synal
MeV SIMS capillary microprobe with secondary electron trigger
USA, Fort Worth, 01.11.2016, CAARI 24 Conference

M. Döbeli
Ion-Solid Interaction for Secondary Ion Mass Spectrometry (SIMS)
Switzerland, Zurich, 14.11.2016, CCMX Advanced Course

O. Fredin, N. Akçar, A. Romundset, R. Reber, S. Ivy-Ochs, P.W. Kubik, D. Tikhomirov, M. Christl, F. Høgaas, C. Schlüchter
New chronological constraints on the deglaciation of northernmost and southernmost Norway based on in-situ cosmogenic nuclides
Norway, Trondheim, 24.05.2015, 4th PAST Gateways International Conference

J. D. Galván, Lukas Wacker, Jan Wunder, Ulf Büntgen
COSMIC: Extraterrestrial evaluation of global-scale tree-ring dating
Germany, Hannover, 03.03.2016, DPG 2016 conference

E. García Morabito, C. Terrizzano, R. Zech, S. Willett, M. Yamin, N. Haghipour, L. Wüthrich, M. Christl, J.M. Cortés, V. Ramos
^{10}Be surface exposure dating reveals strong active deformation in the central Andean backarc interior
Austria, Vienna, 21.04.2016, EGU General Assembly 2016

E. García Morabito, R. Zech, V. Zykina, M. Christl
Surface exposure dating of moraines in the Chagan Uzun Valley, Altai Mountains
Sweden, Stockholm, 09.06.2016, Third Nordic Workshop on cosmogenic nuclide techniques

B. González Domínguez, M. S. Studer, P. A. Niklaus, N. Haghipour, C. McIntyre, L. Wacker, S. Zimmermann, L. Walthert, F. Hagedorn, S. Abiven
Drivers of soil organic matter vulnerability to climate change. Part I: Laboratory incubations of Swiss forest soils and radiocarbon analysis
Austria, Vienna, 17.04.2016, EGU General Assembly 2016

T. Gordijn, K. Hippe, I. Hajdas, S. Ivy-Ochs, V. Picotti, M. Christl
Characterization of sediment storage with cosmogenic nuclides, a study of a fluvial catchment on the Bolivian Altiplano
Switzerland, Geneva, 18.11.2016, Swiss Geoscience Meeting

K. Gückel, M. Christl, T. Shinonaga
Determination Of Plutonium And Americium In Snow On Mt Zugspitze
Finland, Helsinki, 28.09.2016, Ninth International Conference On Nuclear And Radiochemistry - NRC9

K. Guex, A. Ojeda, J. Koch , C. Schneider, M. Döbeli, T. Lippert, D. Günther
Quantitative analysis of La(1-x)CaxMnO3 PLD thin films by UV-fs-LA-ICP-TOF-MS
Switzerland, Beatenberg, 18.-19.11.2016, CH-Analysis

I. Hajdas, U. Sojc, S. Ivy-Ochs, N. Akçar, P. Deline
New radiocarbon chronology of a late Holocene landslide event in the Mont Blanc massif, Italy
Austria, Vienna, 17-22.04.2016, EGU General Assembly 2016

I. Hajdas, M. Maurer, M. Röttig, B. Chevrier, S. Loukou, A. Mayor, M. Rasse , L. Lespez, C. Tribolo, B. Lebrun, A. Camara, E. Huysecom
Radiocarbon Chronology of Human Settlement and paleoenvironment in Senegal, West Africa
Scottland, Edinburgh, 27.06.-01.07.2016, Radiocarbon and Archeaology

I. Hajdas, M. Maurer, M. Röttig
Towards routine ^{14}C dating of mortar at the AMS laboratory ETH Zurich
Scottland, Edinburgh, 27.06.-01.07.2016, Radiocarbon and Archeaology

U. M. Hanke, M. W.I. Schmidt, C. McIntyre, C. M. Reddy, L. Wacker, T. I. Eglinton
Compound specific radiocarbon analyses to apportion sources of combustion products in sedimentary pyrogenic carbon deposits
Austria, Vienna, 17.04.2016, EGU General Assembly 2016

U. M. Hanke, C. P. McIntyre, M. W.I. Schmidt, L. Wacker, T. I. Eglinton
Extraneous carbon assessment in ultra-microscale radiocarbon analysis using benzene polycarboxylic acids (BPCA)
Austria, Vienna, 17.04.2016, EGU General Assembly 2016

D. Michalska, I. Hajdas, D. Sikora, M. Maurer, M. Röttig
Methodological aspect of mortars radiocarbon dating basing on samples from castle from Gora Przemysla, Poznań, Poland
Scottland, Edinburgh, 27.06.-01.07.2016, Radiocarbon and Archeaology

I. Hajdas and participants of MODIS project
Overview of preparation methods applied in the Mortar Dating Intercomparison Study (MODIS)
Scottland, Edinburgh, 27.06.-01.07.2016, Radiocarbon and Archeaology

D. Saracoğlu, I. Hajdas, M. Maurer, M. Röttig, S. Ivy-Ochs, H.A. Synal
Re-visiting radiocarbon ages of Oetzi the Ice Man
Switzerland, Geneva, 19.11.2016, Swiss Geoscience Meeting

F. Gianotti, M. G. Forno, I. Hajdas, G. Monegato, R.Pini, C. Ravazzi
Glacial culmination and decay sequences: new data from a core in the Ivrea end-moraine system (NW Italy)
Austria, Vienna, 17.-22.04.2016, EGU General Assembly 2016

L. Eggenschwiler, I. Hajdas, V. Picotti , P. Cherubini, M. Saurer
Using Stratigraphy and Dendrochronology near Tebano, Senio Northern Apennines to Analyze the Impact of Climatic Fluctuation at Certain Frequencies
Switzerland, Geneva, 19.11.2016, 14th Swiss Geoscience Meeting

B. Giaccio, I. Hajdas, R. Isaia, S. Nomade
Reconciling the time scales of the climatic and cultural processes at the age of the Camapanian Ignimbrite super-eruption (ca. 40 ka BP)
Italy, Bologna, 16.-17.2016, AIQUA

N.Haghipour, T. I. Eglinton, C. McIntyre, J. D. Khatooni, D. Hunziker, A. Mohammadi
Paleo-climate and paleo-environment reconstruction based on a high-resolution, multi-proxy Holocene lake record from Lake Urmia (NW Iran)
Austria, Vienna, 17.04.2016, EGU General Assembly 2016

L. Hendriks, I. Hajdas, M. Küffner, C. McIntyre, N. C. Scherrer, E.S.B. Ferreira
Microscale radiocarbon dating of paintings
Netherlands, The Hague, 11.-13.05.2016, AIA conference

L. Hendriks, I. Hajdas, M. Küffner, E. S.B. Ferreira
Microscale radiocarbon dating from paintings to cultural heritage objects
Belgium, Brussels, 27.-28.10.2016, Relics at the lab workshop

K. Hippe
Advances in Quaternary Geochronology
Italy, Padova, 28.01.2016, Department of Geoscience Seminar

K. Hippe, S. Ivy-Ochs, F. Kober, M. Christl, C. Fogwill, C. Turney, D. Rood, M. Lupker, C. Schlüchter, L. Wacker, R. Wieler
Advances in cosmogenic surface exposure dating: Using combined in situ ^{14}C - ^{10}Be analysis for deglaciation scenarios
Austria, Vienna, 21.04.2016, EGU General Assembly 2016

K. Hippe, A. Fontana, I. Hajdas, S. Ivy-Ochs
A high-resolution ^{14}C chronology from the Cormor alluvial megafan (Tagliamento glacier, NE Italy) for the reconstruction of Alpine glacier activity during 50-20 ka Bè
Austria, Vienna, 21.04.2016, EGU General Assembly 2016

K. Hippe
Advances in Quaternary Geochronology: Examples from combined in situ ^{14}C-^{10}Be analysis
Sweden, Stockholm, 10.06.2016, Third Nordic Workshop on cosmogenic nuclide techniques

S. Ivy-Ochs, C. Wirsig, J. Zasadni, K. Hippe, M. Christl, N. Akçar, C. Schlüchter
Reaching and abandoning the furthest ice extent during the Last Glacial Maximum in the Alps
Austria, Vienna, 21.04.2016, EGU General Assembly 2016

S. Ivy-Ochs, C. Wirsig, J. Zasadni, K. Hippe, M. Christl, N. Akçar, C. Schlüchter
Into and out of the Last Glacial Maximum in the Alps
Switzerland, Geneva, 19.11.2016, Swiss Geoscience Meeting

S. Ivy-Ochs
30 years of Cosmogenic Chronology
Stockholm, Sweden, 08.06.2016, Third Nordic Workshop on cosmogenic nuclide techniques

S. Ivy-Ochs
Update on assessing temporal trends in the timing of Younger Dryas glacier expansions across Europe
Scotland, Inchnadamph, 25.05.2016, Leverhulme Network Meeting: Younger Dryas in Europe

H. Kerschner, A. Moran, S. Ivy-Ochs
Younger Dryas equilibrium line altitudes and precipitation patterns in the Alps
Austria, Vienna, 21.04.2016, EGU General Assembly 2016

O. Kronig, J. M. Reitner, M.Christl, S. Ivy-Ochs
Use of cosmogenic nuclides to date relict rock glaciers
Sweden, Stockholm, 08.06.2016, Third Nordic Workshop on cosmogenic nuclide techniques

O. Kronig, J. M. Reitner, M.Christl, S. Ivy-Ochs
Regional climatic significance of relict rock glaciers in the Eastern Alps
Germany, Potsdam, 22.06.2016, International Conference on Permafrost

O. Kronig, J. M. Reitner, M.Christl, S. Ivy-Ochs
Dating the stabilisation age of relict rockglaciers
Switzerland, Geneva, 18.11.2016, Swiss Geoscience Meeting

M. Lupker, J. Lavé, C. France-Lanord, P. Kubik, M. Christl, D. Bourlès, J. Carcaillet, C. Maden, R. Wieler, M. Rahman, D. Bezbaruah, L. Xiaohan
Downstream lag of cosmogenic-derived denudation in the eastern Himalayan syntaxis
USA, San Francisco, 14.12.2016, AGU conference

A. Madella, R. Delunel, P.H. Blard, N. Akçar, F. Schlunegger, M. Christl
Paleo-erosion rates and paleo-elevation inferred through cosmogenic ^{10}Be and ^{21}Ne in northernmost Chile
Switzerland, Geneva, 19.11.2016, 14th Swiss Geoscience Meeting

X. Maeder, A. Neels, M. Döbeli, A. Dommann, J. Ast, H. Rudigier, B.Widrig, J. Ramm
Synthesis of Al-Ni and Al-Ni-O Coatings by Cathodic Arc Evaporation
USA, San Diego, 25.04.2016, ICMCTF-43

S. Martin, S. Ivy-Ochs, M. Rigo, A. Viganò, V. Alfimov, C. Vockenhuber, K. Hippe
Changing landscape in the Sarca Valley (Trentino) during the Quaternary. Climate or earthquakes?
Italy, Napoli, 07.09.2016, 88th Congresso della Società Geologica Italiana

S. Maxeiner, M. Christl, R. Gruber, A. Müller, M. Suter, H.-A. Synal, C. Vockenhuber
Exploding capillaries and toasted power supplies: status of the 300kV prototype AMS system
Switzerland, Zurich, 16.03.2016, AMS Seminar

S. Maxeiner, H.-A. Synal, M. Christl, M. Suter, A. Müller, C. Vockenhuber
Status of the 300 kV multi isotope AMS project
Germany, Hannover, 03.03.2016, DPG 2016 conference

K.-U. Miltenberger, M. Christl, A. Müller, H.-A. Synal, C.Vockenhuber
Improved Al-26 measurements with absorber setup at low energies
Germany, Hannover, 03.03.2016, DPG 2016 conference

K.-U. Miltenberger, A. Müller, M. Suter, H.-A. Synal, C.f Vockenhuber
^{26}Al measurements using AlO- ions and a gas-filled magnet
Finland, Jyväskylä, 07.07.2016, ECAART 12 conference

K.-U. Miltenberger, M. Schulte-Borchers, A. Müller, M. George, M. Döbeli, H.-A. Synal
MeV-SIMS at ETH Zürich with CHIMP
Italy, Trieste, 27.09.2016, Joint ICTP-IAEA Advanced Workshop on High Sensitivity 2D & 3D
Characterization and Imaging with Ion Beams

K.-U. Miltenberger, M. Schulte-Borchers, A. Müller, M. George, M. Döbeli, H.-A. Synal
MeV-SIMS with the Capillary Heavy Ion MicroProbe
Switzerland, Zurich, 25.11.2016, Zürich PhD Seminar

N. Mozafari Amiri, D. Tikhomirov, Ç. Özkaymak, Ö. Sümer, B. Uzel, S. Ivy-Ochs, C. Vockenhuber,
H. Sözbilir, N. Akçar
Holocene seismic activity of the Yavansu fault, western Turkey
Switzerland, Geneva, 19.11.2016, 14th Swiss Geoscience Meeting

N. Mozafari Amiri, D. Tikhomirov, Ö. Sümer, Ç. Özkaymak, B. Uzel, S. Ivy-Ochs, C. Vockenhuber, H.
Sözbilir, N. Akçar
Determination of paleoseismic activity over a large time-scale: Fault scarp dating with ^{36}Cl
Austria, Vienna, 21.04.2016, EGU General Assembly 2016

N. Mozafari Amiri, Ö. Sümer, D. Tikhomirov, Ç. Özkaymak, B. Uzel, S. Ivy-Ochs, C. Vockenhuber, H.
Sözbilir, N. Akçar
*Using ^{36}Cl fault scarp dating to model Holocene paleoseismic activity in the Büyük Menderes graben,
western Turkey*
Austria, Vienna, 21.04.2016, EGU General Assembly 2016

A. M. Müller, M. Döbeli, H.-A. Synal
High resolution gas ionization chamber in proportional mode
Finland, Jyväskylä, 04.07.2016, ECAART 12 Conference

A. M. Müller, M. Döbeli, M. Christl, S. Maxeiner, H.-A. Synal, C. Vockenhuber
The new ETH 300 kV multi-isotope AMS system
USA, Fort Worth, 31.10.2016, CAARI 24 Conference

M. Oberhänsli, M.Seifert, A.Sindelar, I.Hajdas, R. Turck
Dating Iron age copper mining activities in the Oberhalbstein Valley ((Canton Grisons, CH) using charcoal
Austria, Innsbruck, 21.-25.09.2016, Alpine Copper II. International Workshop

A. Ojeda, C.W. Schneider, M. Döbeli, T. Lippert, A. Wokaun
Plasma plume dynamics and rebounds in pulsed laser deposition: influence of the background
France, Lille, 05.05.2016, EMRS Spring 2016, Symposium C

E. Opyrchał, J. Zasadni, P. Kłapyta, M. Christl, S. Ivy-Ochs
Latest Pleistocene glacial advances in the Veľká Studená Valley (Tatra Mountains, Slovakia)
Sweden, Stockholm, 08.06.2016, Third Nordic Workshop on cosmogenic nuclide techniques

E. Opyrchał, J. Zasadni, P. Kłapyta, M. Christl, S. Ivy-Ochs
Lateglacial glaciations in the High Tatra Mountains, based on [10]Be surface exposure dating
Switzerland, Zurich, 30.11.2016, AMS Seminar

E. Opyrchał, J. Zasadni, P. Kłapyta, M. Christl, S. Ivy-Ochs
Lateglacial deglaciation of the High Tatra Mountains
Switzerland, Geneva, 18.11.2016, Swiss Geoscience Meeting

S. Patra, N. Akçar, S. Ivy-Ochs, G. Monegato
Tracking the pace of Middle Pleistocene Revolution in the southern Alpine Foreland
Switzerland, Into and out of the Last Glacial Maximum in the Alps, 19.11.2016, 14th Swiss Geoscience Meeting

R. Pellitero, B.R. Rea, M. Spagnolo, P. Hughes, R. Braithwaite, H. Renssen, S. Ivy-Ochs, A. Ribolini, J. Bakke, S. Lukas
Glacier-derived climate for the Younger Dryas in Europe
Austria, Vienna, 21.04.2016, EGU General Assembly 2016

R. Pellitero, B.R. Rea, M. Spagnolo, J. Bakke, S. Ivy-Ochs, C. Frew, P. Hughes, A. Ribolini, H. Renssen, S. Lukas
From bed topography to ice thickness: GlaRe, a GIS tool to reconstruct the surface of palaeoglaciers
Austria, Vienna, 21.04.2016, EGU General Assembly 2016

M. Döbeli, A. Dommann, X. Maeder, A. Neels, J. Ramm, H. Rudigier, J. Thomas, B. Widrig
The Influence of Oxygen on the Phase Formation at the Al70Cr30 Target Surface and the Synthesized Coatings in Cathodic Arc Evaporation
USA, San Diego, 27.04.2016, ICMCTF-43

R. Reber, C. Litty, A. Madella, R. Delunel, N. Akçar, M. Christl, F. Schlunegger
Environmental control on erosion along the entire western margin of Peru
Switzerland, Geneva, 19.11.2016, 14th Swiss Geoscience Meeting

T. Sançar, C. Zabcı, N. Akçar, V. Karabacak, M. Yazıcı, H. Akyüz, A. Önal, S. Ivy-Ochs, M. Christl, C. Vockenhuber
Preliminary results on the deformation rates of the Malatya Fault (Malatya-Ovacık Fault Zone, Turkey)
Austria, Vienna, 21.04.2016, EGU General Assembly 2016

A. Sookdeo, L. Wacker, S. Fahrni, C.P. McIntyre, M. Friedrich, F.Reinig, W. Tegel, D. Nievergelt, B. Kromer, U. Büntgen
Speed Dating: a rapid way to determine the radiocarbon age of wood by EA-AMS
Germany, Hannover, 03.03.2016, DPG 2016 conference

A. Sookdeo, L.Wacker, M.Friedrich, F.Reinig, M.Pauly, D. Nievergelt, B. Kromer, G. Helle, U. Büntgen, F. Adolphi, R. Muscheler, J. Beer
One Tree at time: Untangling Variations in ^{14}C during the Younger Dryas
Japan, Yokohama, 01.07.2016, Goldschmidt 2016

G. Steinhauser, T. Niisoe, K. H. Harada, K. Shozugawa, S. Schneider, H.-A. Synal, C. Walther, M. Christl, K. Nanba, H. Ishikawa, A. Koizumi
Resuspension of deposited radioactive material from the Fukushima Daiichi NPP site
Austria, Vienna, 22.04.2016, EGU General Assembly 2016

H.-A. Synal
Exciting Examples of Dating Art Historic Objects: Mona Lisa
Switzerland, Zurich, 14.12.2016, AMS Seminar

H.-A. Synal
"Hunting Atoms" Progressing Accelerator Mass Spectrometry (AMS) for Radiocarbon Dating
Great Britain, Bristol, 28.09.2016, Launching Event of the Bristol AMS Facility

H.-A. Synal
How far can we get? Latest progress in Accelerator Mass Spectrometry (AMS)
Hungary, Budapest, 12.04.2016, RANC-2016

H.-A. Synal
Progress in Accelerator Mass Spectrometry: How far have we travelled? How far can we get?
Romania, Sinaia, 28.06.2016, Carpatian Summer School in Physics

H.-A. Synal
Progress in Accelerator Mass Spectrometry: How far have we travelled? How far can we get?
South Korea, Busan, 13.05.2016, International Particle Accelerator Conference 2016

H.-A. Synal
Wie alt ist Mona Lisa? Atome lügen nicht...
Switzerland, Sursee, 03.11.2016, Öffentlicher Abendvortrag

H.-A. Synal
How old is Mona Lisa? Atoms don't lie...
Switzerland, Zurich, 07.10.2016, Symposium Mona and the x-ray man

H.-A. Synal
"Small is Beautiful" Trends in AMS Facilities
Austria, Vienna, 14.10.2016, Symposium 20 Years of VERA

H.-A. Synal
Welcome on Behalf of the Swiss Vacuum Society
Czech Republic, Praha, 08.11.2016, Pragovac 25

H.-A. Synal
Introduction to the Laboratory of Ion Beam Physics
Switzerland, Zurich, 14.11.2016, CIRP Colooaboration Meeting

C. Terrizzano, R. Zech E. García Morabito, N. Haghipour, M. Christl, J. Likermann, J. Tobal, M. Yamin
Surface exposure dating of moraines and alluvial fans in the Southern Central Andes
Austria, Vienna, 20.04.2016, EGU General Assembly 2016

J. Tobal, E. García Morabito, M. Christl, C. Terrizzano, M. Ghiglione
Deformation in extra-Andean Patagonia at the latitude of the Chile Triple Junction (46º-47º S): tectonic and climatic implications
Switzerland, Bern, 20.09.2016, University of Bern

V. Vanacker, J. Schoonejans, S. Opfergelt, Y. Ameijeiras-Mariño, M. Christl
Kinetically limited weathering at low denudation rates in semi-arid climates
USA, San Francisco, 14.12.2016 , AGU conference

V. Vanacker, J. Schoonejans, Y. Ameijeiras-Mariño, S. Opfergelt, M. Christl
Anthropogenic Disturbances to Geomorphic Processes. New insights from cultural landscapes in the Western Mediterranean.
France, Chambery, 27.06.2016, JGM Conference

V. Vanacker, J. Schoonejans, Y. Ameijeiras-Mariño, S. Opfergelt, M. Christl
Human disturbances to soil systems in the Mediterranean
Belgium, Louvain-laNeuve, 07.07.2016, 13es Journées d'Etude des Sols

V. Vanacker, A. Molina, N. Bellin, M. Christl
Human-induced geomorphic change across environmental gradients
USA, San Francisco, 14.12.2016 , AGU conference

M. Villa-Alfageme, E. Chamizo, M. López-Lora, T. Kenna, N. Casacuberta, M. Christl
MEASUREMENT OF ^{236}U AT THE GEOTRACES EAST PACIFIC ZONAL TRANSECT
Spain, Sevilla, 08.11.2016, II International conference radioecological processes

C. Vivo-Vilches, J.-M. López-Gutiérrez, M. García-León, C. Vockenhuber, T. Walczyk
^{41}Ca detection with Accelerator Mass Spectrometry at low energies: measurements on the 1 MV system at the Centro Nacional de Aceleradores
Spain, Zaragoza, 28.11.2016, CPAN Conference 2016

C. Vockenhuber
^{41}Ca measurements at 500 kV
Germany, Hannover, 03.03.2016, DPG 2016 conference

C. Vockenhuber
Recent developments in AMS at low energies at ETH Zurich
Spain, Seville, 26.10.2016, CNA Seminar

L. Wacker
Progresses in Radiocarbon Dating
Iceland, Reykjavik, 28.04.2016, Arctic drift wood meeting

L. Wacker, U. Büntgen, M. Friedrich, R. Friedrich, J.D. Galván, I. Hajdas, B. Kromer, F. Miyake,
D. Nievergelt, F. Reinig, A. Sookde, H.-A. Synal, T. Westphal
Benefits of a new calibration curve with a high temporal resolution
Scottland, Edinburgh, 30.06.2016, Radiocarbon and Archeaology

C. Welte, L. Wacker, B. Hattendorf, M. Christl, J. Koch, C. Yeman, J. Fohlmeister, S. F. M. Breitenbach,
A. H. Andrews, L. Robinson, J. R. Farmer, D. Günther, H.-A. Synal

Laser Ablation – Accelerator Mass Spectrometry: rapid and spatially resolved radiocarbon analyses of carbonate archives
Finland, Jyväskylä, 04.07.2016, ECAART 12 Conference

D. Wiedemeier, N. Haghipour, C. P. McIntyre, T.I. Eglinton, M. W. I. Schmidt
Tracing pyrogenic carbon suspended in rivers on a global scale
Austria, Vienna, 17.04.2016, EGU General Assembly 2016

H. Wittmann, F. v. Blanckenburg, N. Dannhaus, J. Bouchez, J. Gaillardet, J.L. Guyot, L. Maurice, H. Roig, N. Filizola, M. Christl
Denudation and weathering rates from meteoric $^{10}Be/^9Be$ ratios in the Amazon basin
USA, San Francisco, 13.12.2016, AGU conference

C. Yeman, C. Welte, B. Hattendorf, J. Koch, M. Christl, L. Wacker, A. H. Andrews, A. Freiwald
Continuous ^{14}C analysis of marine carbonates by laser ablation coupled with AMS
USA, Portland ME, 07.06.2016, 4th International Sclerochronology Conference

S. Yeşilyurt, U. Doğan, N. Akçar, V. Yavuz, S. Ivy-Ochs, C. Vockenhuber, F. Schlunegger, C. Schlüchter
Late Quaternary Glaciations of Kavuşşahap Mountains
Switzerland, Geneva, 19.11.2016, 14th Swiss Geoscience Meeting

S. Yeşilyurt, N. Akçar, U. Doğan, V. Yavuz, S. Ivy-Ochs, C. Vockenhuber, F. Schlunegger, C. Schlüchter
Extensive Quaternary glaciations in eastern Turkey
Austria, Vienna, 21.04.2016, EGU General Assembly 2016

M. Yu, T. I. Eglinton , N. Haghipour, D. B. Montluçon, L.Wacker, P.i Hou, M. Zhao
Influence of hydrodynamic sorting on the composition and age of Yellow River suspended particulate organic matter
USA, San francisco, 12.12.2016, AGU conference

SEMINAR
'CURRENT TOPICS IN ACCELERATOR MASS SPEKTRO-METRY AND RELATED APPLICATIONS'

Spring semester

24.02.2016
Martina Schulte Borchers (ETHZ), MeV SIMS with a capillary microprobe for molecular imaging

02.03.2016
Fusa Miyake (Nagoya Univ.), Past cosmic ray intensity study by using cosmogenic nuclides

09.03.2016
Maria Fischer (EMPA), Aluminum Oxynitride Thin Films Deposited by Reactive Direct Current Magnetron Sputtering

16.03.2016
Sascha Maxeiner (ETHZ), Exploding capillaries and toasted power supplies: status of the 300kV prototype AMS system

22.03.2016
Caroline Welte (ETHZ), On the progress of radiocarbon measurements using Laser Ablation Accelerator Mass Spectrometry (LA-AMS)

23.03.2016
Stefano Casale (University of Pisa), Dating the Late Glacial in the Rhaetian Alps

30.03.2016
Klaus-Ulrich Miltenberger (ETHZ), Improving AMS measurements of ^{26}Al

06.04.2016
Nuria Casacuberta (ETHZ), ^{236}U, Pu-isotopes and ^{129}I in the ocean: an update

13.04.2016
Sean Francis Gallen (ETHZ), What controls the spatial and temporal distribution of erosion rate in dead orogens? A case-study of landscape evolution in the Southern Appalachians, USA

20.04.2016
Carlos Vivo Vilches (CNA), ^{41}Ca AMS at low energies: applications at ETHZ and CNA Sevilla

27.04.2016
Anne Claude (Uni Bern), Swiss forleland scenery: evolution during the Quaternary

04.05.2016
Naoto Ishikawa (JAMSTEC), Application of bulk and compound-specific radiocarbon analyses to ecological research in aquatic systems

11.05.2016
Alfred Priller (Uni Wien), Implementation of an Ion Laser Interaction Setup (ILIAS) into the Vienna Environmental Research Accelerator (VERA)

18.05.2016
Alessandro Fontana (Univ. Padua), From the Dolomites to the Adriatic: Late-Quaternary evolution of Venetian-Friulian Plain (NE Italy)

25.05.2016
Ruslan Cusnir (IRA CHUV), Bioavailability of Plutonium in the karstic freshwater environments

01.06.2016
Maxi Castrillejo (UAB), ^{129}I and ^{236}U along the GEOVIDE transect in the North Atlantic Ocean

Fall semester

20.09.2016
Kevin Norton (Victoria Univ. Wellington, NZ), From A(ntarctica) to Z(urich): Wellington's contribution to cosmogenic nuclides as glacial timekeepers

21.09.2016
Blanca Ausin (ETHZ), Evaluating the (a)synchroncity of climatic signals in marine records from the Shackleton sites

28.09.2016
Nasim Mozafari Amiri (Univ. Bern), Determination of Holocene major seismic activities in the Western Anatolia using cosmogenic ^{36}Cl

05.10.2016
Daniele De Maria (ETHZ), Development of a beam profile monitor for phase space measurements

26.10.2016
Alfio Vigano (Univ. Trento), Monitoring present seismicity and reconstructing past seismicity in north-eastern Italy

02.11.2016
Andreas Türler (PSI), Theranostics: image and treat using combinations of non-standard radionuclides

09.11.2016
Adam Sookdeo (ETHZ), Untangeling radiocarbon variations during the Younger Dryas climate event

16.11.2016
Thomas Walczyk (NUS, Singapore), Evaluation of calcium and strontium metabolism in animals and humans using isotopic techniques

23.11.2016
Chia-Yu Chen (ETHZ), Spatial patterns of denudation rates in the southern Central Range of Taiwan

30.11.2016
Ewelina Opyrchał (AGH, Krakow, Poland), Lateglacial glaciations in the High Tatra Mountains, based on ^{10}Be surface exposure dating

07.12.2016
Christiane Yeman (ETHZ), Laser Ablation-radiocarbon-AMS: first data from the revised setup

14.12.2016
Hans-Arno Synal (ETHZ), A review on the involvement of LIP in authenticity studies of Mona Lisa

14.12.2016
Pascal Cotte (Lumiere Technology), New discoveries on the Mona Lisa. How science can change the History

21.12.2016
Sahra Talamo (MPI Leipzig), Neanderthals meet anatomically Modern Human? Radiocarbon vis-a-vis Rolex clock

THESES (INTERNAL)

Term papers/Bachelor

Tanja Frei
Reconstructing the Sennis and Malun Paleoglaciers and understanding the early Alpine Lateglacial
ETH Zurich (Switzerland)

Nicolas Gay
What happened in Bösbächi?
ETH Zurich (Switzerland)

Monika Isler
Zur Methodik der Radiokarnondatierung
University of Zurich (Switzerland)

Diploma/Master theses

Jochem Braakhekke
The Last Glacial Maximum around Lago d'Orta, Italy
ETH Zurich (Switzerland)

Daniele De Maria
Beam Profile Monitor, Design and measurements of the phase space
ETH Zurich (Switzerland)

Catharina Dieleman
Alluvial Fan Development and Landscape Evolution in the Region of Sedrun (Surselva, Kanton Graubünden, Switzerland)
ETH Zurich (Switzerland)

Loren Eggenschwiler
Using dendrochronology and stratigraphy to understand climatic events near Tebano, Italy
University of Zurich

Tiemen Gordijn
Sediment storage and its effect on cosmogenic nuclides, a study of a fluvial catchment on the Bolivian Altiplano
ETH Zurich (Switzerland)

Klaus-Ulrich Miltenberger
Improving AMS measurements of ^{26}Al
ETH Zurich (Switzerland)

Doctoral theses

Franziska Lechleitner
Characterizing the legacy of carbon in karst systems: isotopic and chronological applications on stalagmites

ETH Zurich (Switzerland)
Sascha Maxeiner
Improving Designs of future AMS facilities
ETH Zurich (Switzerland)

Bao Rui
Sedimentological control on organic carbon burial in shallow marginal seas
ETH Zurich (Switzerland)

Martina Schulte-Borchers
MeV SIMS based on a capillary microprobe for molecular imaging
ETH Zurich (Switzerland)

THESES (EXTERNAL)

Diploma/Master theses

Lorena Burckhart
Die früh- und hochmittelalterlichen Funde und Befunde von Sogn Pieder in Domat/Ems
University of Zurich (Switzerland)

Melissa Graber
^{10}Be surface exposure dating of the Rhone- and the Aare-glacier
University of Bern (Switzerland)

Nicole Meichtry
Last Deglaciation of the Aare Valley
University of Bern (Switzerland)

Katja Mettler
When did the accumulation stop in the Northern Alpine Foreland at the end of the Last Glacial Cycle?
University of Bern (Switzerland)

Annina Ruppli
Erosion in an upland Mediterranean environment
University of Zurich (Switzerland)

Doctoral theses

Victor Alarcon
Development of Charged Particle Detection Systems for Materials Analysis With Rapid Ion Beams
Université Pierre et Marie Curie, Paris (France)

Benjamin Campforts
What goes up must come down: Improving numerical simulation of landscape evolution at different timescales
Université catholique de Louvain (Belgium)

Anne Claude
Landscape evolution of the northern Alpine Foreland: constructing a temporal framework for early to middle Pleistocene glaciations
University of Bern (Switzerland)

Reto Grischott
Spatially and temporally variable catchment-wide denudation rates - clues from the Alps
ETH Zurich (Switzerland)

Giulia Guidobaldi
Late Pleistocene glaciers of Northern Apennines as archive for paleoclimatic reconstruction in the Mediterranean basin
University of Pisa (Italy)

Ursina Jecklin-Tischhauser
Die Kirchenanlage Sogn Murezi in Tomils (GR) vom frühen bis ins späte Mittelalter
University of Zürich (Switzerland)

Timo Jäger
Transparent Conductive Oxides by Magnetron Sputtering for Solar Energy Applications
ETH Zurich (Switzerland)

Alejandro Ojeda
Physical processes in pulsed laser deposition
ETH Zurich (Switzerland)

Markus Pichler
Photocatalytically active lanthanum titanium oxynitrides: an investigation of structural electronic and photoelectrochemical properties
ETH Zurich (Switzerland)

Michael Reinke
Surface Kinetics of Titanium Isopropoxide in Chemical Vapor Deposition of Titanium Dioxide and Barium Titanate
EPF Lausanne (Switzerland)

Jerome Schoonejans
Constraining soil formation and development through quantitative analyses of chemical weathering and physical erosion
Université catholique de Louvain (Belgium)

Kenta Shimamoto
Strain-engineering of PLD-grown Orthorhombic Rare-earth Manganate Thin Films
ETH Zurich (Switzerland)

Barbara Zollinger
Alpine permafrost and its effect on chemical weathering,soil organic carbon pools and time-split soil erosion - A case study from the Eastern Swiss Alps
University of Zurich (Switzerland)

COLLABORATIONS

Australia

The University of Western Australia, Oceans Institute, Crawley

Deakin University, Institute for Frontier Materials, Geelong

The Australian National University, Department of Nuclear Physics, Canberra

Austria

AlpS - Zentrum für Naturgefahren- und Riskomanagement GmbH, Geology and Mass Movements, Innsbruck

Geological Survey of Austria, Sediment Geology, Vienna

University of Innsbruck, Institute of Geography, Geology and Botany, Innsbruck

University of Salzburg, Geography and Geology, Salzburg

University of Vienna, VERA, Faculty of Physics, Vienna

Vienna University of Technology, Institute for Geology, Vienna

Belgium

Royal Institute for Cultural Heritage, Brussels

Université catholique de Louvain, Earth and Life Institute, Louvain-la-Neuve

Canada

Chalk River Laboratories, Dosimetry Services, Chalk River

University of Ottawa, Department of Earth Sciences, Ottawa

China

China Institute for Radiation Protection, Dosimetry Services, Taiyuan city

Chinese Academy of Sciences, Institute of Botany, Beijing

Oceanographic Institution, Qingdao

Peking University, Accelerator Mass Spectrometry Lab., Beijing

University of Science and Technology, Materials Science, Beijing

Denmark

Danfysik A/S, Taastrup

Technical University of Denmark, Department of Photonics Engineering, Roskilde

Univ. Southern Denmark, Department of Physics, Chemistry and Pharmacy, Odense

Finnland

University of Jyväskylä, Physics Department, Jyväskylä

France

Aix-Marseille University, Collège de France, Aix-en-Provence

Commissariat à l'énergie atomique et aux énergies alternatives, Laboratoire des Sciences du Climat et de l'Environnement (LSCE), Gif-sur-Yvette Cedex

Laboratoire de biogeochimie moléculaire, Strasbourg

Laboratoire des sciences du climat et de l'environnement (LSCE), CNRS-CEA-UVSQ, Gif-sur-Yvette

Université de Savoie, Laboratoire EDYTEM, Le Bourget du Lac

Université Pierre et Marie Curie, Ion Beam Laboratory, Paris

Germany

Alfred Wegener Institute of Polar and Marine Research, Marine Geochemistry, Bremerhaven

Bundesamt für Strahlenschutz, Strahlenschutz und Umwelt, Neuherberg

Deutsches Bergbau Museum, Bochum

GFZ German Research Centre for Geosciences, Earth Surface Geochemistry and Dendrochronology Laboratory, Potsdam

Helmholtz-Zentrum Dresden-Rossendorf, DREAMS, Rossendorf

Helmholtz-Zentrum München, Institut für Strahlenschutz, Neuherberg

Hydroisotop GmbH, Schweitenkirchen

Leibniz-Institut für Ostseeforschung Warnemünde, Marine Geologie, Rostock

LMU-Munich, Geosciences, Munich

Marum, Micropalaeontology - Paleoceanography and Marine Seimentologie, Bremen

Regierungspräsidium Stuttgart, Landesamt für Denkmalpflege, Esslingen

Reiss-Engelhorn-Museen, Curt-Engelhorn-Zentrum Archäometrie gGmbH, Mannheim

ROWO AG, Herbholzheim

University of Applied Sciences, TH Köln, Technology Arts Sciences, Clogne

University of Bochum, Geology, Bochum

University of Cologne, Physics Department and Institute of Geology and Mineralogy, Cologne

University of Hannover, Institute for Radiation Protection and Radioecology, Geosciences, Hannover

University of Heidelberg, Institute of Environmental Physics, Heidelberg

University of Hohenheim, Institute of Botany, Stuttgart

University of Münster, Institute of Geology and Paleontology, Münster

University of Tübingen, Department of Geosciences, Tübingen

Hungary

Hungarian Academy of Science, Institute of Nuclear Research (ATOMKI), Debrecen

India

Inter-University Accelerator Center, Accelerator Division, New Dehli

Italy

CAEN S.p.A., Viareggio

CNR Rome, Institute of Geology, Rome

Geological Survey of the Provincia Autonoma di Trento, Landslide Monitoring, Trento

INGV Istituto Nazionale di Geofisica e Vulcanologia, Sez. Sismologia e Tettonofisica, Rome

University of Bologna, Deptartment Earth Sciences, Bologna

University of Padua, Department of Geosciences, Geology and Geophysics, Padua

University of Pisa, Department of Geology, Pisa

University of Salento, Department of Physics, Lecce

University of Turin, Department of Geology, Turin

Japan

University of Tokai, Department of Marine Biology, Tokai

Liechtenstein

OC Oerlikon AG, Balzers

Oerlikon Surface Solutions AG, Balzers

Netherlands

NIOZ Royal Netherlands Institute for Sea Research, Coastal Systems Sciences, Texel

New Zealand

University of Waikato, Radiocarbon Dating Laboratory, Waikato

Victoria University of Wellington, School of Geography, Environment and Earth Sciences, Wellington

Norway

Norwegian Geological Survey, Trondheim

Norwegian University of Science and Technology, Physical Geography, Trondheim

Rogaland Fylkeskommune, Savanger

University of Bergen, Department of Earth Science, Research Climate and Biology, Bergen

University of Norway, The Bjerkness Centre for Climate Res., Bergen

Poland

Adam Mickiewicz University, Department of Geology, Poznan

University of Marie Curie Sklodowska, Department of Geography, Lublin

Romania

Horia Hulubei - National Institute for Physics and Nuclear Engineering, Magurele

Singapore

National University of Singapore, Department of Chemistry, Singapore

Slovakia

Comenius University, Faculty of Mathematics, Physics and Infomatics, Bratislava

Slovenia

Geological Survey of Slovenia, Ljubljana

South Korea

KATRI Korea Apparel Testing and Research Institute, Seoul

Spain

University of Murcia, Department of Plant Biology, Murcia

University of Seville, Physics Department and National Center for Accelerators, Seville

Sweden

Lund University, Department of Earth and Ecosystem Sciences, Lund

University of Uppsala, Angström Institute, Upsalla

Switzerland

ABB Ltd, Baden

ABB Ltd, Lenzburg

Bern University of Applied Sciences, Hochschule der Kunst, Bern

Centre Hospitalier Universitaire Vaudois, Institut de radiophysique, Lausanne

Dendrolabor Wallis, Brig

EAWAG, Department of atmospheric sceince, Zurich

Empa, Research Groups: Nanoscale Materials Science, Mechanics of Materials and Nanostructures, X-ray Analytics, Functional Polymers and Hochleistungskeramik, Thin Films, Corrosion and Joining Technology, Dübendorf

ENSI, Brugg

EPFL, Microengineering, Lausanne

ETH Zurich, Zurich
Department of Earth Sciences: Geologiacal Institute, Institute of Geochemistry and Petrology, Institute of Isotope Geochemistry and Mineral Resources, Engineering Geology
Department of Chemistry and Applied Biosciences: Laboratory of Inorganic Chmeistry, Trace Element and Micro Analysis
Department of Materials: Polymer Technology, Institute of Metals Research
Department of Environmental Systems Science: Environmental Physics
Department of Physics: Laboratory for Solid State Physics

Geneva Fine Art Analysis Sarl, Lancy, Geneva

Gübelin Gem Lab Ltd. (GGL), Luzern

Helmut Fischer AG, Hünenberg

Kanton Bern, Achäologischer Dienst, Bern

Kanton Graubünden, Kantonsarchäologie, Chur

Kanton Solothurn, Kantonsarchäologie, Solothurn

Kanton St. Gallen, Kantonsarchäologie, St. Gallen

Kanton Turgau, Kantonsarchäologie, Frauenfeld

Kanton Zug, Kantonsarchäologie, Zug

Kanton Zürich, Kantonsarchäologie, Dübendorf

Labor für quartäre Hölzer, Affoltern a. Albis

Laboratiore Romand de Dendrochronologie, Cudrefin

Landesmuseum, Zurich

NAGRA, Wettingen

Office et Musée d'Archéologie Neuchatel, Neuchatel

Paul Scherrer Institut (PSI), Laboratories for Micro and Nanotechnology, for Atmospheric Chemistry, for Radiochemistry and Environmental Chemistry, Materials Group, Radiochemistry, Villigen

Research Station Agroscope Reckenholz-Tänikon ART, Air Pollution / Climate Group, Zurich

Stadt Zürich, Amt für Städtebau, Zurich

SUPSI, Dipartimento ambiente costruzioni e design (DACD), Lugano

Swiss Federal Institute for Forest, Snow and Landscape Reseach (WSL), Landscape Dynamics, Dendroecology and Soil Sciences, Birmensdorf

Swiss Federal Institute of Aquatic Science and Technology (Eawag), SURF, Dübendorf

Swiss Gemmological Institute, SSEF, Basel

Swiss Institute for Art Research, SIK ISEA, Zurich

University of Basel, Departement Altertumswissenschaften und Institut für Prähistorische und Naturwissenschaftliche Archäologie (IPNA), Basel

University of Bern, Department of Chemie and Biochemistry, Climate and Environmental Physics, Oeschger Center for Climate Research and Institute of Geology and Geography, Bern

University of Freiburg, Faculty of Environmentat and Natural Resources, Department of Physics ,Freiburg

University of Geneva, Department of Anthropology and Ecology, Geology and Paleontology, and Quantum Matter Physics, Geneva

University of Lausanne, Department of Geology, Lausanne

University of Zurich, Institute of Geography, Abteilung Ur- und Frühgeschichte, Institut für Evolutionäre Medizin , Zurich

Turkey

Dokuz Eylül University, Department of Geological Engineering, Izmir

Istanbul Technical University, Faculty of Mines, Istanbul

Tunceli Üniversitesi, Geology Department, Tunceli

United Kingdom

Brithish Arctic Survey, Cambridge

Durham University, Department of Geography, Durham

Newcastle University, School for History, Classics and Archaeology, Newcastle

Queen Mary University of London, School of Geography, London

University of Aberdeen, School of Geosciences, Aberdeen

University of Bristol, School of Chemistry and School of Earth Sciences, Bristol

University of Oxford, Department of Earth Sciences, Oxford

USA

Colorado State University, Department of Environmental and Radiological Health Sciences, Fort Collins

Columbia University, LDEO, New York

Cornell University, Department of Earth and Atomospheric Sciences, New York

Idaho National Laboratory, National and Homeland Security, Idaho Falls

Lamont-Doherty Earth Observatory, Department of Geochemistry, Palisades

NOAA Fischeries, Pacific Islands Fisheries Science Center, Honolulu

University of Utah, Geology and Geophysics, Salt Lake City

Woods Hole Oceanographic Institution, Center for Marine and Environmental Radioactivity, Marine Chemistry and Geochemistry, Woods Hole

VISITORS AT THE LABORATORY

Carlos Vivo-Vilches
University of Seville, Centro Nacional de Aceleradores CAN, Sevilla, Spain
10.01.2016 - 08.05.2016

Sara Rubinetti
University of Turin, Turin, Italy
18.01.2016 - 07.02.2016

Jiro Matsuo
University of Kyoto, Quantum Science and Engineering Center, Kyoto, Japan
21.01.2016

Froidevaux Pascal
CHUV, Lausanne, Switzerland
29.01.2016

Silvana Martin
University of Padova, Department of Geosciences, Padova, Italy
01.02.2016 - 05.02.2016

Ewelina Opyrchal
AGH University of Science and Technology, Department of Environmental Analysis, Cartography and
Economic Geology, Krakow, Poland
08.02.2016 - 31.03.2016

Maxi Castrillejo
Universitat Autonoma de Barcelona, Barcelona, Spain
10.02.2016 - 31.01.2017

Fusa Miyake
Nagoya University, Nagoya, Japan
22.02.2016 - 06.03.2016

Djibril Faye
Universidde de Lisboa, Instituto Superior Técnico IST, Lisboa, Spain
07.03.2016 - 21.03.2016

Freddy Reing
Eidg. Forschungsanstalt WS, Brimensdorf, Switzerland
04.04.2016 - 13.04.2016

Maren Pauly
Deutsches GeoForschungsZentrum, Potsdam, Germany
04.04.2016 - 13.04.2016

Heinz Gäggeler
Paul Scherrer Institut PSI, Villigen, Switzerland
26.04.2016

Stefano Casale
University of Florence and University of Pisa, Florence/Pisa, Italy
01.05.2016 - 30.07.2016

Benjamin Lehmann
Université de Lausanne Insitut des Dynamiques de la Surface Terrestre, Lausanne, Switzerland
15.05.2016 - 15.06.2016

Ewelina Opyrchal
AGH University of Science and Technology, Department of Environmental Analysis, Cartography and Economic Geology, Krakow, Poland
23.05.2016 - 10.07.2016

Milko Jakšić
Ruđer Bošković Institute, Zagreb, Kroatia
15.06.2016 - 18.06.2016

Katharina Gückel
Helmholtz Zentrum München, Institut für Strahlenschutz (ISS), Arbeitsgruppe Experimentelle Radioökologie, Munich, Germany
12.07.2016 - 13.07.2016

Silvana Martin
University of Padova, Department of Geosciences, Padova, Italy
02.08.2016 - 04.08.2016

Marc Ostermann
University of Innsbruck, Geology Department, Innsbruck, Austria
19.08.2016 - 26.08.2016

Bryan Lougheed
Uppsala University, Department of Earth Sciences, Uppsala, Sweden
19.09.2016 - 23.09.2016

Helen Flewlass
Max Planck Institut for Evolutionary Anthropology, Leipzig, Germany
26.09.2016 - 30.09.2016

Bernd Kromer
Max Planck Institut for Evolutionary Anthropology, Leipzig, Germany
26.09.2016 - 30.09.2016

Satinath Gargari
Inter-University Accelerator Centre, Neu Delhi, India
17.10.2016 - 10.11.2016

Rajveer Sharma
Inter-University Accelerator Centre, Neu Delhi, India
17.10.2016 - 10.11.2016

Raimund Muscheler
Lund University, Lund, Sweden
01.11.2016

Xiongxin Dai
China Institute for Radiation Protection Taiyuan City, Shanxi, China
14.11.2016 - 15.11.2016

Xueqi Chang
China Institute for Radiation Protection Taiyuan City, Shanxi, China
14.11.2016 - 15.11.2016

Anxi Cui
China Institute for Radiation Protection Taiyuan City, Shanxi, China
14.11.2016 - 15.11.2016

Liye Liu
China Institute for Radiation Protection Taiyuan City, Shanxi, China
14.11.2016 - 15.11.2016

Changming Deng
China Institute for Radiation Protection Taiyuan City, Shanxi, China
14.11.2016 - 15.11.2016

Wei Cheng
China Institute for Radiation Protection Taiyuan City, Shanxi, China
14.11.2016 - 15.11.2016

Franziska Slotta
Deutsches GeoForschungsZentrum, Potsdam, Germany
14.11.2016 - 18.11.2016

Cotte Pascal
Lumiere Technology, Paris, France
13.12.2016

TRAINEES AND STUDENTS AT THE LABORATORY

Klaus-Ulrich Miltenberger, Master student
ETH Zurich, Zurich, Switzerland
14.09.2015 - 21.03.2016

Maurizio Hitz, Trainee
Kantonsschule Wettingen, Wettingen, Switzerland
22.02.2016 - 11.03.2016

Karl Seidler, Apprentice
ETH Zurich, Zurich, Switzerland
01.03.2016 - 30.09.2016

Daniele De Maria, Master student
ETH Zurich, Zurich, Switzerland
07.03.2016 - 30.09.2016

Duygu Saracoglu, Trainee
Istanbul Technical University, Istanbul, Turkey
24.08.2016 - 16.09.2016

Devta Sekhar, Trainee
Kantonsschule Olten, Olten, Switzerland
26.09.2016 - 30.09.2016

Stefan Schellinger, Trainee
Eidg. Forschungsanstalt WSL, Birmensdorf, Switzerland
26.10.2016 - 01.11.2016